Lynn Winspear

# ESSENTIALS
## Edexcel
## INTERNATIONAL GCSE
## Biology

# Contents

## The Nature and Variety of Living Organisms

- 4 Characteristics of Living Organisms
- 5 Variety of Living Organisms
- 8 Exam Practice Questions

## Structures and Functions in Living Organisms

- 10 Levels of Organisation
- 13 Biological Molecules
- 16 Movement of Substances In and Out of Cells
- 19 Nutrition in Plants
- 23 Nutrition in Humans
- 28 Respiration
- 30 Gas Exchange in Plants
- 32 Gas Exchange in Humans
- 36 Transport in Plants
- 40 Transport in Humans
- 45 Excretion
- 48 Coordination and Response in Humans
- 50 Coordination and Response in Plants
- 51 Nervous Coordination in Humans
- 54 Exam Practice Questions

## Reproduction and Inheritance

- 58 Reproduction
- 59 Human Reproduction
- 60 Plant Reproduction
- 63 Inheritance
- 72 Exam Practice Questions

# Contents

## Ecology and the Environment

- **74** The Organism in the Environment
- **76** Feeding Relationships
- **78** Cycles Within Ecosystems
- **80** Human Influences on the Environment
- **82** Exam Practice Questions

## Use of Biological Resources

- **84** Food Production – Crop Plants
- **87** Food Production – Microorganisms
- **90** Food Production – Fish
- **91** Selective Breeding
- **92** Genetic Modification (Genetic Engineering)
- **94** Cloning
- **96** Exam Practice Questions

- **100** Answers
- **106** Glossary of Key Words
- **112** Index

# Characteristics of Living Organisms

## Eight Characteristics of Life

Every living organism shows all eight characteristics at some time during its life.

**Movement:** Animals and plants move in **response to external stimuli**. An animal may move its whole body, whereas parts of a plant can move. Some single cells move around using cilia or flagellum or pseudopodia. Cell structures (organelles) also move in the cytoplasm.

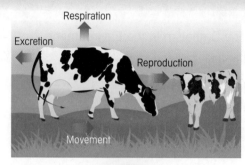

**Respiration:** Living organisms **release energy** from nutrients.

**Sensitivity:** Living organisms **detect and respond to changes** in their surroundings and to changes inside themselves. For example, plant shoots grow towards light.

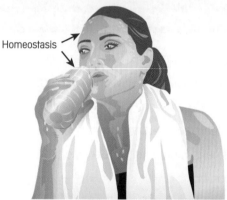

**Homeostasis:** Animals, plants and single-celled organisms **control their internal conditions**. For example, they control how much water they take in and get rid of so that the amount of water inside them stays almost constant.

**Growth:** All living organisms can **permanently increase their mass**. Their cells increase in number and size.

**Reproduction:** They produce offspring.

**Excretion:** Living organisms **get rid of materials that are the waste products** from reactions taking place inside their cells, for example carbon dioxide waste from respiration.

**Nutrition:** Living organisms take in and use nutrients as raw materials **to build cells and to release energy**.

## Quick Test

1. Which characteristic of living organisms describes the release of energy from nutrients?
2. State two ways in which living organisms grow.
3. Use one example to explain what we mean by homeostasis.
4. Suggest one stimulus which a part of a plant can respond to.
5. Describe two ways in which a single living cell can move.

**Key Words** — Movement • Respiration • Sensitivity • Homeostasis • Growth • Reproduction • Excretion • Nutrition

# Variety of Living Organisms

## The Five Kingdoms

Living organisms are classified (grouped) into five kingdoms. **Organisms in each kingdom have similar characteristics.**

**Viruses** are not cells; they are DNA or RNA surrounded by protein. They cannot carry out all eight characteristics of life and so they do not fit into the table of five kingdoms.

| Kingdom | Features | Feeding Method | Movement |
|---|---|---|---|
| Protoctista | Single-celled, have a nucleus, some have some chloroplasts, usually no cell wall, organelles present | Photosynthesis, or ingestion of other organisms, or both | Move using cilia or flagella |
| Bacteria | Single-celled, no nucleus, no chloroplasts, have a cell wall | Absorb nutrients through cell wall, or produce their own | May or may not move |
| Fungi | Multicellular, have a nucleus, no chloroplasts, have cell walls | Acquire nutrients from decaying material | No mechanisms for movement |
| Plants | Multicellular, have a nucleus, chloroplasts and cell walls | Require sunlight to make food through photosynthesis | Most don't move |
| Animals | Multicellular, have a nucleus, no chloroplasts, no cell walls | Acquire nutrients by ingestion | Most animal groups use muscles to move |

## Plants

Plants are **multicellular** organisms. Their cells contain chloroplasts and are able to carry out photosynthesis. Their cells have cellulose cell walls and they store carbohydrates as starch or sucrose. Examples include flowering plants, such as a cereal, (e.g. maize) and herbaceous legumes (e.g. peas or beans).

Plant cells have:
- a permanent **vacuole** filled with cell sap
- a **cell wall** made of cellulose
- a **nucleus**, which controls the cell activities
- **cytoplasm**, where most chemical reactions take place
- a **cell membrane**, which controls the passage of substances in and out of the cell
- **mitochondria**, where most energy is released in **respiration**
- **ribosomes**, where protein synthesis takes place.

Some plant cells also have **chloroplasts**, which contain chlorophyll that makes them green. **Chlorophyll absorbs light energy**, which the plant cell uses in photosynthesis to make food.

Plants store carbohydrates in their cells as sucrose or starch.

Bean (herbaceous legume)    Maize (cereal)

## Animals

Animals are **multicellular**. All animals have the following features in common:
- Animal cells **do not** have cell walls, permanent vacuoles, chloroplasts or carry out photosynthesis.
- Animals **can** move around from place to place, usually use nerves to coordinate their responses to stimuli, and store carbohydrates as glycogen.

**Key Words**: Virus • Vacuole • Cell wall • Nucleus • Cytoplasm • Cell membrane • Mitochondria • Ribosome • Chloroplast

# Variety of Living Organisms

## Fungi

Fungi cannot carry out photosynthesis. Some (e.g. yeast) are single cells.

Most are made of **fine threads called hyphae**. Fungi may store carbohydrates as **glycogen**.

Fungi **secrete digestive enzymes** onto food (**extra-cellular digestion**) and then absorb the products of digestion. This is called **saprotrophic nutrition**.

Hyphae:
- have **cell walls made of chitin**
- contain many nuclei
- are grouped together to make **networks and structures called a mycelium** (e.g. Mucor)
- do not contain chloroplasts and so **cannot photosynthesise**.

**Cell Structure of Yeast – Single-celled Fungus**

Nucleus – contains the DNA that carries the genetic code for making the enzymes needed in respiration

Cytoplasm – where proteins, including enzymes used in anaerobic respiration, are made

Cell wall – made of chitin

Mitochondria – where aerobic respiration occurs

Cell membrane – allows gases and water to enter and leave the cell freely while acting as a barrier to other, larger chemicals

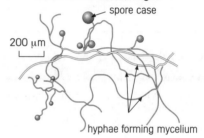

**Mucor – Multi-celled Fungus**

spore case

200 μm

hyphae forming mycelium

## Bacteria

Bacteria are **microscopic single-celled organisms**. Most bacteria feed on living or dead organisms, but some can photosynthesise.

All bacteria have:
- cell walls that are **not made of cellulose**
- cell membranes
- cytoplasm
- **plasmids** – tiny loops of DNA
- a **single circular chromosome** made of DNA shaped into a large loop
- **no nucleus**.

*Lactobacillus bulgaricus* is a rod-shaped bacterium. It is used to make yoghurt from milk.

*Pneumococcus* is a spherical bacterium that causes pneumonia.

**Cell Structure of Bacteria**

Cytoplasm – where proteins, including enzymes used in aerobic and anaerobic respiration, are made

Circular DNA – the DNA floats freely in the cytoplasm

Plasmid – a tiny loop of DNA

Cell membrane – allows gases and water to enter and leave the cell freely while acting as a barrier to other, larger chemicals

Cell wall – gives the bacterial cell strength

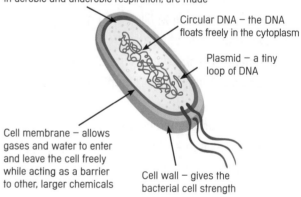

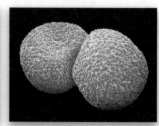

Lactobacillus bulgaricus

Pneumococcus

**Key Words:** Hyphae • Glycogen • Enzyme • Extra-cellular digestion • Saprotrophic nutrition • Chitin • Mycelium • Plasmid

# Variety of Living Organisms

## Protoctists

**Protoctists** are organisms that are microscopic and single-celled.

Some protoctists are like animal cells, others are like plant cells. Some cause diseases. **Plasmodium** is a parasite that is carried by mosquitoes. It causes malaria in humans by infecting and destroying red blood cells.

**Amoeba** is a large protoctist, like a free-living animal cell. **Chlorella** is a single celled organism that moves about. It has a chloroplast and so can photosynthesise like a plant.

## Viruses

Viruses are not cells; they **are particles of DNA or RNA surrounded by protein**. Viruses:
- are **smaller than bacteria**
- can only **reproduce inside living cells**
- are many sizes and shapes.

When a virus invades a living cell:
1. The virus attaches to a host cell and injects its genetic material into the cell.
2. It uses the cell to make components for a new virus.
3. The host cell splits open to release the virus.

The **HIV** virus causes AIDS. The virus enters white blood cells and destroys them. White blood cells are part of the body's immune response. A person with the HIV virus can no longer easily destroy common bacteria and viruses that they catch, for example, the common cold virus.

The **tobacco mosaic virus** prevents tobacco leaf cells making chloroplasts, so the plant cannot photosynthesise well.

**Virus**
- Strand of genetic material
- Protein coat

**HIV Virus**
- Protein coat
- Genetic material

**Tobacco Mosaic Virus**

## Pathogens

Some fungi, bacteria and protoctists are **pathogens**. All viruses are pathogens.

## Quick Test

1. Name the five kingdoms that living organisms are grouped into.
2. What is the difference between the way carbohydrates are stored in plants and in animals?
3. Explain how the tobacco mosaic virus causes poor growth in tobacco plants.
4. Explain why people with the HIV virus may become very ill when they catch a cold.

**Key Words** — Pathogen

# Exam Practice Questions

**1** Scientists classify organisms into groups that have similar features. Match each statement to the correct group by inserting the correct letters in the table. [5]

| Animals | D |
| --- | --- |
| Plants | B |
| Bacteria | E |
| Fungi | C |
| Protoctists | A |

A: The organism is a single cell, swims in a pond, and has a chloroplast.

B: This organism stores carbohydrates as starch and its cells have cellulose walls.

C: The organism has cells with walls and it stores carbohydrates as glycogen.

D: The cells of this organism store carbohydrates as glycogen but don't have chitin.

E: This single-celled organism has a circular chromosome but no nucleus.

**2** Describe two structural differences between a plant cell and a fungal cell. [2]

*Fungi do not have Chloroplasts and cannot photosynthesise*
*Plant cell walls have Cellulose, Fungi have Chitin.*

**3** Some fungi are made of fine threads that are clumped into networks. [2]

Name:

a) the fine fungal threads

*Hyphae*

b) the network formed by the clumped threads.

*Mycelium*

**4** Describe how a fungus obtains its nutrients. [3]

*Secrete enzymes on decaying material.*

# Exam Practice Questions

**5** Complete this table with examples of pathogens and diseases in each group. [4]

| Group | Organism | Disease |
|---|---|---|
| Bacteria ~~Pathogen~~ ✗ | Pneumococcus | Pneumonia |
| Virus | HIV | Aids |
| Protoctista | Plasmodium | Malaria |

**6** People have made use of the organism called *Lactobacillus bulgaricus*. Describe *Lactobacillus bulgaricus* and state how people make use of it. [3]

Answer: Bacterium, rod shaped, used to make yoghurt.

**7 a)** Label the diagram of the plant cell. [6]

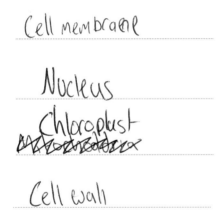

Cell membrane
Nucleus
Chloroplast
~~Mitochondria~~
Cell wall

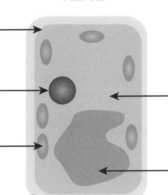

~~Vacuole~~ Cytoplasm
Vacuole

**b)** Which three labelled features are not found in animal cells? [3]

Chloroplast, cell wall, Vacuole ✓ ✓ ✓

16 Marks

# Levels of Organisation

## Cell Structures

Living organisms are made of cells. All cells have a **cell membrane** which surrounds the **cytoplasm**. Cytoplasm is a jelly-like substance that contains enzymes. Many chemical reactions take place in the cytoplasm.

Cell **organelles** are found in the cytoplasm. There are different types of organelle and each type carries out a specific cell function. The nucleus, mitochondria, ribosomes and chloroplasts are examples of organelles.

## Animal and Plant Cells

Most animal cells and plant cells have the following parts:
- **Nucleus** – contains the genetic material of the cell and controls cell activities. It is surrounded by its own membrane.
- **Cytoplasm** – most chemical reactions take place here.
- A **cell membrane** – controls the passage of substances in and out of the cell.
- **Mitochondria** – where most energy is released in **respiration**.
- **Ribosomes** – where protein synthesis occurs.

Chemical reactions inside cells are controlled by **enzymes** found in the **cytoplasm** and **mitochondria**.

**Plant cells** also have the following parts:
- A **cell wall** – (made out of cellulose) used to strengthen the cell. (Algae also have a cell wall.)
- **Chloroplasts** – absorb light energy to make food.
- A **permanent vacuole** – filled with cell sap that helps to support the cell.

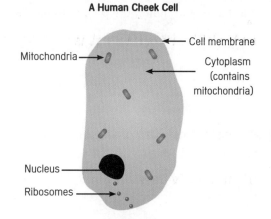

A Human Cheek Cell

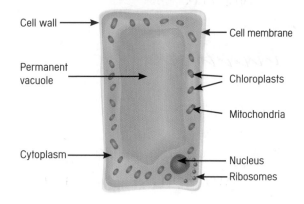

A Plant Cell from a Leaf

## Specialised Cells

Cells may be specialised to carry out a particular job.

| Root Hair Cells | Ovum (egg cell) | Xylem | White Blood Cells | Sperm Cells | Palisade Cells | Red Blood Cells | Neurone |
|---|---|---|---|---|---|---|---|
| Tiny hair-like extensions that increase the surface area of the cell for absorption. | Large cell that carries food reserves for the developing embryo. | Long, thin, hollow cells used to transport water through the stem and root. | Can change shape in order to engulf and destroy invading microorganisms. | Have a tail, which allows them to move. | Packed with chloroplasts for photosynthesis. | No nucleus, so packed full of haemoglobin to absorb oxygen. | Long, slender axons that can carry nerve impulses. |

**Key Words**        Organelle

# Levels of Organisation

## Tissues

Large multicellular organisms, like humans, develop **systems for exchanging materials**. As the organism develops, cells **differentiate** so that they can carry out different jobs.

A **tissue** is a group of cells that have a **similar structure and function**. For example:
- **muscle tissue contracts** so we can move
- **glandular tissue produces substances** such as enzymes and hormones
- **epithelial tissue covers** organs.

**Muscle Tissue**
Can contract to bring about movement

**Glandular Tissue**
Can produce substances such as enzymes and hormones

**Epithelial Tissue**
Covers all parts of the body

## Organs

**Organs** are made of **tissues**. One organ may contain several tissues.

For example, the **stomach** is an organ that contains:
- **muscle tissue** that contracts to churn the contents
- **glandular tissue** to produce digestive juices
- **epithelial tissue** to cover the outside and inside of the stomach.

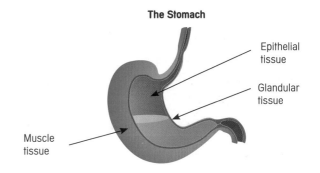

The Stomach

## Organ systems

**Organ systems** are **groups** of organs that carry out a particular function.

For example, the **digestive system** includes:
- glands, such as the **pancreas** and **salivary glands**, that produce digestive juices
- the **stomach** and **small intestine**, where digestion takes place
- the **liver**, which produces **bile** to help break down fats
- the **small intestine**, where the **soluble food** is **absorbed** into the blood
- the **large intestine**, where **water is absorbed** from undigested food, leaving faeces.

Humans also have other organ systems, such as the excretory system involving the kidneys, and the reproductive system.

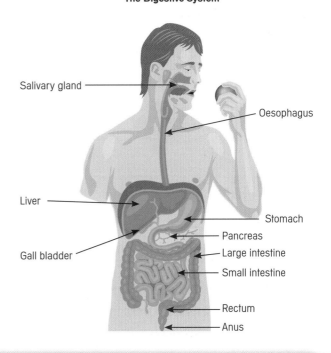

The Digestive System

**Key Words**: Differentiate • Tissue • Organ

# Levels of Organisation

## Plant Organs

Plant organs include:
- stems
- roots
- leaves
- flowers.

**Plant tissues** include:
- **epidermal tissues** that cover the plant
- **mesophyll**, where photosynthesis takes place
- **xylem** and **phloem** that transport substances around the plant.

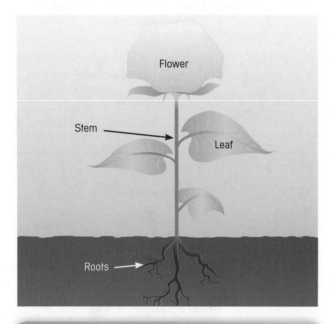

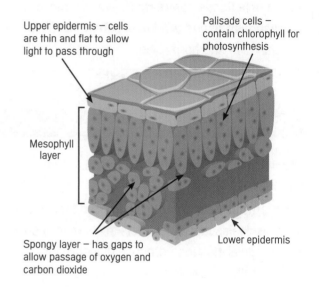

Cross Section of a Leaf

## Understanding Levels of Organisation

The following flow charts should help you to understand the scale and size of cells, tissues and organ systems.

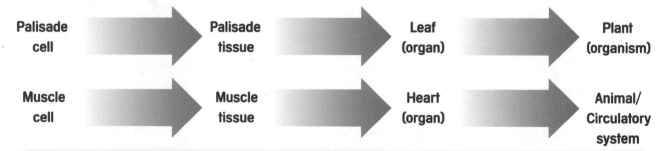

## Quick Test

1. Name the organelles where energy is released in respiration.
2. List three structures found in plant cells that are not found in animal cells.
3. What are a group of cells that have a similar structure and function called?
4. Is the heart a tissue, organ or organ system?
5. Which animal tissue covers the outside and inside of organs?

**Key Words**  Xylem • Phloem

# Biological Molecules

## Elements in the Body

Ninety-six percent of a human's body mass is made of four elements. These are carbon (18%), hydrogen (10%), nitrogen (3%) and oxygen (65%). There are small amounts of other elements too. You use the elements to build molecules that make up cells, and you use some for chemical reactions like **respiration**. Most of the molecules in living organisms are **carbohydrates**, **lipids** or **proteins**. You get these molecules from food.

Carbohydrates, lipids and proteins are long, complicated molecules **made up from small basic units**.

**Carbohydrates** are made from carbon, hydrogen and oxygen arranged to form **sugars like glucose and sucrose**. The sugars are joined together to make **long chains of starch or glycogen**.

**Proteins** are made up of **long chains of amino acids**. Amino acids all contain carbon, hydrogen, oxygen and nitrogen.

**Lipids** are built from **fatty acids and glycerol**. They contain carbon, hydrogen and oxygen.

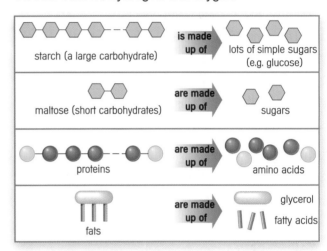

## Testing Foods for Glucose and Starch

**Testing foods for glucose using Benedict's Reagent.**

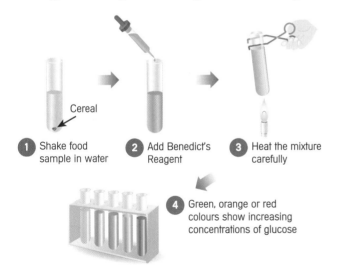

1. Shake food sample in water
2. Add Benedict's Reagent
3. Heat the mixture carefully
4. Green, orange or red colours show increasing concentrations of glucose

**Testing foods for starch using iodine.**

The brown iodine turns blue-black if starch is present in the food

## Enzymes

**Enzymes** are **proteins** that act as **biological catalysts**. They speed up chemical reactions, including those that take place in living cells, e.g. respiration, photosynthesis and protein synthesis.

Enzymes are highly specific. Each one will only speed up a **particular** reaction. Enzyme activity, and therefore the rate of a reaction, can be affected by changes in **temperature** or **pH** level.

**Key Words**: Carbohydrate • Lipid • Protein • Benedict's Reagent • Iodine • Catalyst

# Biological Molecules

## Enzyme Activity and Temperature

Each **enzyme** has an **active site** that only a specific reactant can fit into (**like a key in a lock**). High temperatures stop the 'lock and key' mechanism working.

When enzyme molecules are exposed to high temperatures, the following occurs:

1. The bonds holding the shape of the protein break.
2. The shape of the enzyme's active site is **denatured** (changed irreversibly).
3. The 'lock and key' mechanism no longer works.

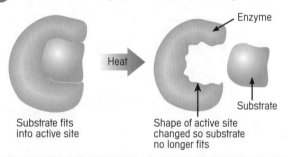

The graph shows the effect of temperature on enzyme activity.

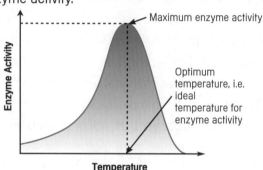

A rise in temperature increases the **frequency of collisions** between substrates and enzymes, and will increase the enzyme activity until the **optimum temperature** is reached. At lower temperatures there are lower collision rates, slowing down the rate of the reaction.

Temperatures above the optimum denature the enzyme.

## Enzyme Activity and pH

The graph shows how changes in pH affect enzyme activity. There is an **optimum pH** at which the enzyme works best. As the pH increases or decreases, the enzyme becomes less and less effective.

If the pH of the solution is outside the optimum, the shape of the enzyme is affected because the bonds holding the enzyme molecule into a precise shape are affected by pH. This changes the shape of the active site and makes it more difficult for the substrate to fit. Changes in pH also affect the way in which the substrate bonds to the active site.

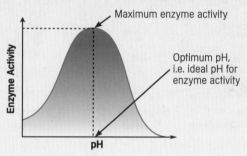

## Quick Test

1. Name the four most common elements found in the body.
2. What are the units that make up: **a)** large carbohydrates **b)** lipids **c)** proteins?
3. Describe how you would test a cereal containing glucose with Benedict's Reagent and what you would see.
4. What is an enzyme?
5. Explain why an enzyme will only act on a specific reactant (substrate).
6. Explain why the rate of a reaction increases as temperature increases, up to the optimum temperature.
7. Describe what is meant by the term 'denatured'.

**Key Words**: Active site • Denatured • Optimum temperature • Optimum pH

# Biological Molecules

## Investigating the Effect of Temperature on Enzyme Activity

To compare how fast enzymes act at different temperatures you can either measure how fast the product is formed, or how fast the substrate is used up.

### How Fast is Oxygen Produced at Different Temperatures?

**Catalase** is an enzyme that **breaks down hydrogen peroxide into oxygen and water**. Catalase is found in living tissues, e.g. potatoes. To see how temperature affects the activity of catalase, you can measure how fast the catalase breaks hydrogen peroxide down to form oxygen at different temperatures.

1. Add potato to the hydrogen peroxide and start the timer.
2. Oxygen is released and makes froth rise up the tube.
3. After 30 seconds, **measure the height of the froth**. The height of froth in 30 seconds is a measure of the **activity of the catalase enzyme**.
4. Repeat the experiment using water baths and test tubes at different temperatures to see how temperature affects the rate at which oxygen bubbles are produced.

It is important to keep the mass and shape of the potato disc constant, and use the same potato to make the discs. You also have to keep the volume of hydrogen peroxide constant and use the same size of test tube.

In this experiment the **independent variable** is **the temperature** and the **dependent variable** is **the height of the froth after 30 seconds**.

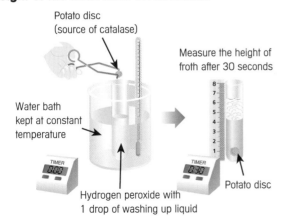

### How Fast is Starch Used Up at Different Temperatures?

**Amylase** is an enzyme that **breaks down starch into glucose**. You can see if starch is present by testing it with iodine (a blue-black colour indicates that starch is present).

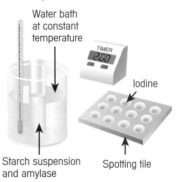

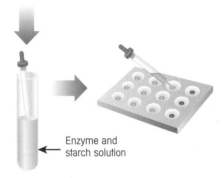

1. Mix together amylase enzyme and starch and keep at a constant temperature.
2. Every 30 seconds remove a drop of the mixture and add to the iodine in the tile.
3. When the amylase has broken down all the starch, the iodine will stay brown.
4. Record **the time it takes for the starch to disappear**.
5. Repeat the experiment at a range of different temperatures.

You must keep the volume and concentration of starch suspension and amylase the same for each temperature.

In this experiment the **independent variable** is **the temperature** of the water bath. The **dependent variable** is **the time taken for the starch to disappear**.

**Key Words**: Catalase • Independent variable • Dependent variable

# Movement of Substances In and Out of Cells

## Movement of Substances

### Diffusion

Substances move in and out of cell membranes by **diffusion**. **Diffusion** is the **net (overall)** movement of the particles of a substance from a region of **high concentration** to a region of **low concentration**.

The **rate** of **diffusion into or out of a cell** is increased when:
- there is a greater surface area of the cell membrane
- there is a greater difference between concentrations inside and outside the cell (a steeper **concentration gradient**)
- the distance the particles have to travel is short
- temperature is increased – this makes the diffusing particles move about faster.

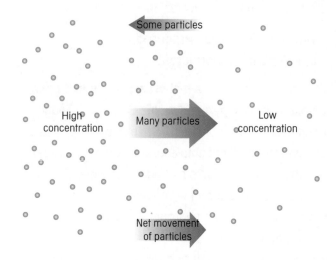

### Active Transport

Substances sometimes need to be absorbed from an area of **low** concentration to an area of **high concentration**, i.e. against a concentration gradient.

This is called **active transport** and it requires **energy** from **respiration**. Plants absorb mineral ions through their root hairs by active transport.

### Osmosis

**Osmosis** is a special type of diffusion involving water molecules.

Osmosis is the **diffusion** of water **from a high concentration of water** (dilute solution) **to a low concentration of water** (concentrated solution) through a **partially permeable membrane** (a membrane that allows the passage of water molecules but not solute molecules).

### Net Movement

In osmosis, the water particles move randomly, colliding with each other and passing through the membrane in both directions. But the **net movement** of molecules is from the area of high water concentration to the area of low water concentration.

You can **predict the direction** of water movement if you know what the **concentration** of the water is. Remember, solute molecules can't pass through the membrane; only the water molecules can.

Movement of water is always from **high** to **low** water concentration.

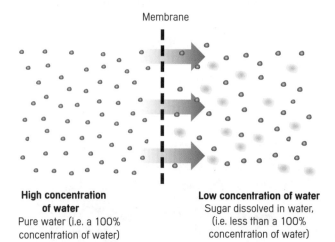

**Key Words**: Diffusion • Concentration gradient • Active transport • Osmosis

# Movement of Substances In and Out of Cells

## Osmosis and Support

**Plant cells** have **inelastic cell walls** that, together with the **water inside the cells**, are essential for the **support** of young non-woody plants.

The cell wall:
- prevents cells from bursting due to excess water
- contributes to rigidity.

The pressure of the water pushing against the cell wall is called **turgor pressure**. A lack of water can cause plants to **droop (wilt)**. As the amount of water inside the cells reduces, the cells become less rigid due to reduced turgor pressure.

As water moves into plant cells **by osmosis**, the **pressure inside the cells increases**. The inelastic cell walls can withstand the pressure and the cells becomes very turgid (rigid). When all the cells are fully turgid, the plant is firm and upright. But, if water is in short supply, cells will start to lose water **by osmosis**. They lose turgor pressure and become flaccid (not rigid), and the plant begins to wilt.

When cells lose a lot of water, the inside of the cells contracts. This is called plasmolysis.

Rigid Plant

Wilting Plant

## Diffusion Experiment in a Non-Living System

Visking tubing is an artificial **partially permeable membrane**. You can knot the end of a length of tubing and fill the bag you have made with a glucose solution. Then tie the top with thread and suspend the bag in a beaker full of water.

After an hour you can **test the water in the beaker for glucose** using Benedict's Reagent. The glucose will have diffused through the visking tubing into the beaker **down its concentration gradient**.

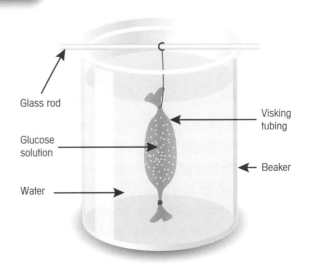

**Key Words**  Turgid • Flaccid • Plasmolysis

# Movement of Substances In and Out of Cells

## Osmosis Experiment in a Non-Living System

To show osmosis taking place in a non-living system, set up a visking tubing bag attached to a glass tube, as shown in the illustration. Fill the bag with sucrose solution. Sucrose is a large molecule that is too big to pass though the visking tubing. Put the bag in a beaker of water and wait for a few hours.

**The liquid slowly rises up the tube.** This is because the concentration of water in the beaker is greater than the concentration of water in the bag and so **water passes through the semi-permeable visking tubing into the bag by osmosis.**

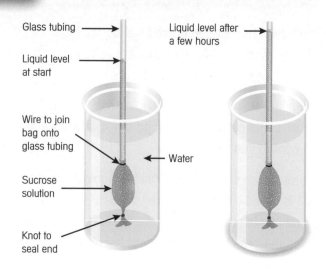

## Osmosis Experiment in a Living System

To show osmosis taking place in a living system, put sticks of potato, all the same length and width at the start, in beakers of water or concentrated sugar solution, as shown in the diagram. **Measure the potato sticks** before you put them in the liquids. After several hours measure them again and **compare the lengths**.

The potato cells that contain more concentrated sugar solution than the surrounding liquid will **gain water by osmosis and therefore increase in length**.

Potato cells that contain less concentrated sugar solution than the surrounding liquid will **lose water by osmosis and shrink in size**.

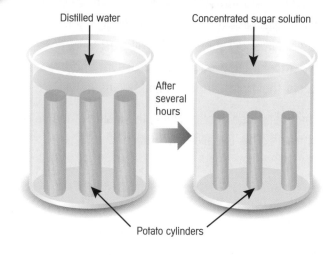

## Quick Test

1. State two differences between diffusion and active transport.
2. State three factors that increase the rate of diffusion.
3. What is meant by the term osmosis?
4. **P2** Explain how osmosis is involved in plant support.
5. **P2** Suggest why a plant begins to wilt if it is flooded by salty water.
6. Explain what will happen to the length of a carrot that is soaked in distilled water for an hour.

# Nutrition in Plants

## Photosynthesis

Green plants don't absorb food from the soil. They make their own food using sunlight. This process is called **photosynthesis**.

Photosynthesis occurs in the cells of **green plants** and **algae** that are exposed to **light**.

During photosynthesis, light energy is absorbed by green **chlorophyll**, which is found in **chloroplasts** in some plant cells and algae.

The four things needed for photosynthesis are:
- **light** from the Sun
- **carbon dioxide** from the air
- **water** from the soil
- **chlorophyll** in the leaves.

The **light energy** is used to convert **carbon dioxide** from the air and **water** from the soil into **sugar** (glucose). **Oxygen** is released as a by-product.

Photosynthesis converts light energy into chemical potential energy. The energy is stored inside the molecules of glucose that are produced.

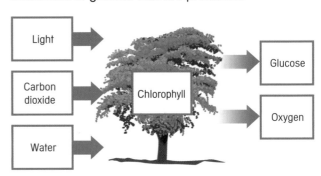

The word equation for photosynthesis is:

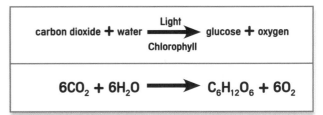

$$6CO_2 + 6H_2O \longrightarrow C_6H_{12}O_6 + 6O_2$$

## Factors Affecting Photosynthesis

There are several factors that may at any time **limit** the rate of photosynthesis.

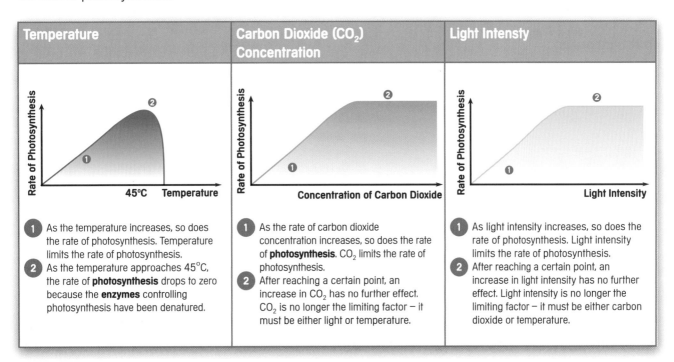

| Temperature | Carbon Dioxide (CO$_2$) Concentration | Light Intensty |
|---|---|---|
| ① As the temperature increases, so does the rate of photosynthesis. Temperature limits the rate of photosynthesis.<br>② As the temperature approaches 45°C, the rate of **photosynthesis** drops to zero because the **enzymes** controlling photosynthesis have been denatured. | ① As the rate of carbon dioxide concentration increases, so does the rate of **photosynthesis**. CO$_2$ limits the rate of photosynthesis.<br>② After reaching a certain point, an increase in CO$_2$ has no further effect. CO$_2$ is no longer the limiting factor – it must be either light or temperature. | ① As light intensity increases, so does the rate of photosynthesis. Light intensity limits the rate of photosynthesis.<br>② After reaching a certain point, an increase in light intensity has no further effect. Light intensity is no longer the limiting factor – it must be either carbon dioxide or temperature. |

**Key Words**: Photosynthesis • Chlorophyll

# Nutrition in Plants

## Plant Leaves

**Photosynthesis** occurs mainly in the leaves of plants. Leaves are specially adapted for efficiency.

For example, a leaf:
- contains the pigment **chlorophyll (which absorbs light)** in millions of chloroplasts, plus other pigments to absorb light from different parts of the spectrum
- is **broad** and **flat** to provide a **huge surface area** to absorb sunlight
- has a **network of vascular bundles** for **support**, and to **transport** water to the cells and remove the products of photosynthesis, i.e. glucose
- has a **thin structure** so the gases (carbon dioxide and oxygen) only have a short distance to travel to and from the cells
- has **stomata** (tiny pores) on the underside of the leaf to allow the **exchange** of **gases**; these are opened and closed by **guard cells**.

During photosynthesis **carbon dioxide** diffuses in through the **stomata** (leaf pores) and **oxygen** diffuses out through the **stomata**. Water is absorbed through the roots.

A leaf has four distinct layers: the **upper epidermis**, the **palisade layer**, the **spongy mesophyll** and the **lower epidermis**.

In a typical leaf:
- the **upper epidermis** is **transparent** to allow sunlight through to the layer below
- the **cells** in the **palisade layer** are near the **top** of the leaf and are packed with **chloroplasts** so they can absorb the maximum amount of light
- the **spongy mesophyll** contains lots of **air spaces** connected to the stomata to allow the optimum exchange of gases.

This internal structure provides a **very large surface area to volume ratio** for efficient gaseous exchange.

Plant cells contain many chloroplasts and are long so they can absorb lots of light. Chloroplasts are not found in all plant cells, for example, root cells don't have chloroplasts as they don't receive any light.

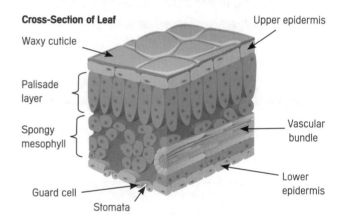

Cross-Section of Leaf

## Quick Test

1. Write the chemical equation for photosynthesis.
2. Describe and explain how the rate of photosynthesis is affected by temperature.
3. Describe how gases enter and leave a leaf.
4. How is the internal surface area to volume ratio of a leaf made very large?
5. Explain how the broad, thin shape of a leaf adapts the leaf for photosynthesis.

**Key Words**  Stomata

# Nutrition in Plants

## Essential Minerals

**Essential minerals** are needed to keep plants healthy and growing properly. Plants absorb dissolved minerals in the soil through their roots.

The minerals are **naturally present** in the soil, although usually in quite **low concentrations**. So farmers use **fertilisers** containing essential minerals to make sure that plants get all the minerals they need to grow.

Each mineral is needed for a different purpose:
- **Nitrates** $NO_3^{2-}$ – used to make amino acids that form proteins for cell growth.
- **Magnesium** $Mg^{2+}$ – used to make the chlorophyll for photosynthesis.

If one or more of the essential minerals is missing (deficient) from the soil, the growth of the plant will be affected.

Experiments can be carried out to see how removing one mineral affects the plants. You do this by growing plants in a soil-less culture. The minerals are then carefully controlled and changed.

## Photosynthesis Experiments

To find out if a plant has been photosynthesising, you can test its leaves for the starch that it stores.

### Testing Leaves for Starch

1. Hold a leaf in tweezers and **dip it in boiling water**. This kills the cells and stops the leaf using up any starch it has stored. Then take the beaker off the heat and turn off the Bunsen burner.
2. Put the leaf into a test tube of **ethanol** and stand the tube in the beaker of boiling hot water. **The ethanol removes the green colour** (chlorophyll) from the leaf.
3. Rinse the leaf in cold water to soften it, spread it on a tile and add iodine solution. If starch is present, the iodine will turn blue-black.

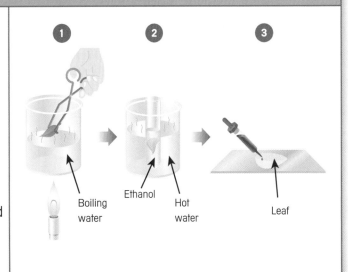

### Plants Need Chlorophyll for Photosynthesis

**Variegated leaves** have only got chlorophyll in the parts that are green. Test a variegated leaf for starch and **only the parts that were green will turn blue-black**.

This shows that the plant can only photosynthesise in the parts of the leaf that contain chlorophyll.

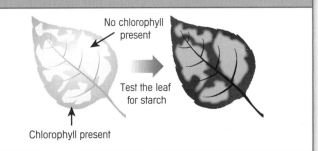

**Key Words**: Fertiliser • Variegated leaf

# Nutrition in Plants

## Photosynthesis Experiments (Cont.)

### Plants Need Light for Photosynthesis

The apparatus can be set up as shown in the diagram.

Starch is not found in the part of the leaf that was shaded, but **starch is found in the unshaded parts** of the leaf. This shows that plants need light for photosynthesis.

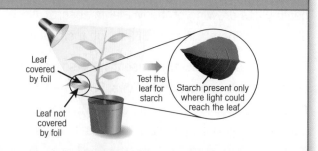

### Plants Need Carbon Dioxide for Photosynthesis

Put a well-watered plant inside a sealed bell jar that contains soda lime. Soda-lime absorbs the carbon dioxide in the jar. After two days in the light, test the leaves for starch. They do not turn iodine black. This means the plant used up its store of starch and has not been able to replace it by photosynthesising, because there is no carbon dioxide in the jar. To check that the plant can make starch when it has carbon dioxide, take the soda-lime out of the jar and test the leaves again for starch after two more days.

### Plants Produce Oxygen During Photosynthesis

When pondweed is illuminated, **bubbles of gas are given off**. The gas can be collected using the apparatus shown. When the tube is full of gas, test for oxygen by putting a glowing splint quickly into the tube.

If the splint relights, oxygen is present. (If you keep the apparatus in the dark, you may be able to collect a gas, but this gas will not relight a glowing splint because it is not oxygen.)

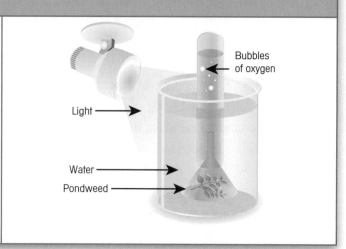

## Quick Test

1. Name two mineral ions needed by plants.
2. Which of these mineral ions is needed by the plant to make **a)** protein? **b)** chlorophyll?
3. **a)** What is a variegated leaf? **b)** How can a variegated leaf tell you about photosynthesis?
4. Describe how you would show that oxygen is given off during photosynthesis.

# Nutrition in Humans

## Nutrients and their Functions

Food is a mixture of different nutrients. Each nutrient has a particular function in the body. Different foods are rich in different nutrients, so you need to eat a range of foods to get all the nutrients you need.

| Nutrient | | Food Source | Function |
|---|---|---|---|
| Carbohydrates (sugar and starch) | | Pasta, rice, cake, bread, potatoes | Provide energy |
| Lipids (fats, oils) | | Oily fish, cheese, butter | Provide energy and act as a store of energy that also insulates the body |
| Protein | | Meat, fish | For growth and repair and as an emergency source of energy |
| Minerals | Calcium | Dairy products | Hardens bones and teeth |
| | Iron | Red meat | Used to make haemoglobin in blood |
| Vitamins | A | Liver, carrots | For healthy hair, skin and eyesight |
| | C | Oranges | Prevents scurvy |
| | D | Eggs | Needed so the body can absorb calcium |
| Dietary fibre | | Wholemeal bread, vegetables and fruit | Keeps food moving easily through the gut |
| Water | | Food and drink | To keep the correct balance of water in the body and replace water lost through sweating, breathing and passing urine |

## A Balanced Diet

Your diet is balanced when the food you eat supplies you with all the nutrients you need **in the correct proportions**.

If you do not eat enough of a particular nutrient you will suffer from a **deficiency disease**. For example, if you do not get enough calcium you may suffer from rickets. A lack of protein will lead to poor growth.

Different people need different amounts of energy-containing nutrients. The amount of energy that a person needs depends on:

- **age** – children and teenagers **need more energy because they are growing** and are very active
- **pregnancy** – pregnant women **need extra energy** to feed the developing baby and carry the extra weight
- **activity level** – **active people** and those who do active jobs **need more energy** than people who are not very active.

To keep your body mass healthy, you must balance the amount of energy in the food you eat with the amount of energy you use up. A diet that supplies more energy than a person uses will make them obese because the excess energy is stored as fat. A diet with insufficient energy will make the person **underweight**.

**Key Words**: Obese

# Nutrition in Humans

## The Digestive System

The alimentary canal is a tube that runs through your body from your mouth to your anus. As food is pushed through it, it is digested and absorbed into the blood.

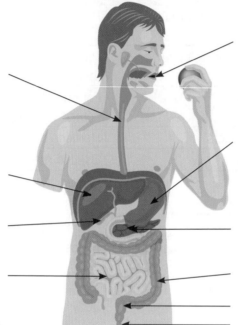

**The Digestive System**

**Mouth**
a) Teeth and tongue break food into small pieces
b) Salivary glands secrete carbohydrase enzymes
   **Digestion of carbohydrate** begins

**Oesophagus**
Muscles in the walls push food down to the stomach by peristalsis

**Stomach**
a) Churns food
b) Secretes **acid that kills bacteria** in food
c) Secretes protease enzyme
   **Protein digestion begins**

**Liver**
Produces **bile**

**Gall bladder**
**Stores bile** and releases it into the small intestine

**Pancreas**
**Produces protease, lipase and carbohydrase enzymes** and releases them into the small intestine

**Small intestine**
a) Produces proteases, lipases, and carbohydrases
   **Digestion of proteins, fats and carbohydrates is completed**
b) **Digested nutrients are absorbed** into the blood

**Large intestine**
**Absorbs water** from the food

**Rectum**

**Anus**

## Moving Food Along

Food is **pushed through the alimentary canal by contractions of muscles** in the walls of the canal. The squeezing action of the gut that pushes food along is called **peristalsis.**

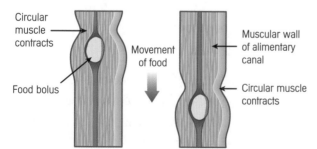

Circular muscle contracts
Food bolus
Movement of food
Muscular wall of alimentary canal
Circular muscle contracts

### Quick Test

1. List seven nutrients that your body requires.
2. Name two foods that are good sources of:
   a) protein
   b) energy
3. **P2** Explain what happens when a person eats more energy-containing food than the body uses.
4. Which parts of the digestive system produces:
   a) bile?
   b) carbohydrase enzymes?
   c) acid?

**Key Words**   **Peristalsis**

# Nutrition in Humans

## Stages of Digestion

Digestion occurs in five main stages.

1. **Ingestion** – when you put food or drink into your mouth.
2. **Digestion** – breaking down large, insoluble nutrient molecules into small, soluble ones that can be absorbed:
   - **Mechanical digestion** is where teeth, muscles and bile break up lumps of food into a fine mush that has a **large surface area for enzymes to act** on.
   - **Chemical digestion** is caused by enzymes. It is the **breakdown of large molecules into small ones** that can be absorbed.
3. **Absorption** – small digested nutrient molecules **move from the alimentary canal through the intestine wall and into the blood**. Digested food is absorbed in the small intestine; water is absorbed in the large intestine.

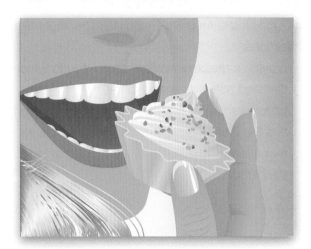

4. **Assimilation** – the digested nutrients travel to the cells in the blood. There they may be used to become part of the cell. For example, amino acids may be used by the cell to make cell proteins. The **nutrients are assimilated**.
5. **Egestion** – undigested food and excess water leave the body through the anus. This is called egestion.

## Action of Bile

**Bile** is produced in the liver and stored in the gall bladder. When it is released into the small intestine, it has two functions:

1. **Bile neutralises the acid** added to food in the stomach. It produces alkaline conditions in which the small intestine enzymes can work well.

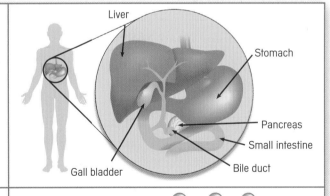

2. **Bile emulsifies lipids**. This makes the lipids form into tiny droplets that have a **very large surface area** for lipase enzymes to act on.

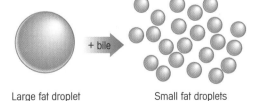

Large fat droplet    Small fat droplets

**Key Words**: Ingestion • Digestion • Mechanical digestion • Chemical digestion • Absorption • Assimilation • Egestion • Bile

# Nutrition in Humans

## How Villi Help

The small intestine is lined with **villi**. The finger-like projections:
- provide a **massive surface area** through which digested nutrients are absorbed
- have a very thin layer of cells between the food and the blood capillary inside them, so it is easy for nutrients to diffuse the short distance to the blood
- are well supplied with blood, which carries absorbed nutrients quickly away from each villus. There is therefore always a steep concentration gradient for more absorption to take place.

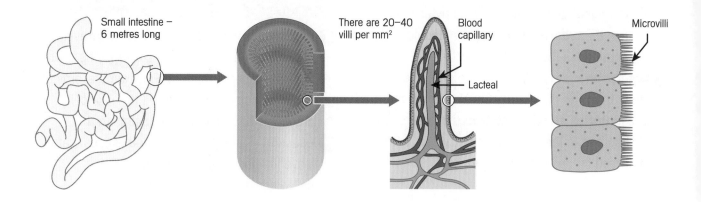

## Action of Enzymes

Digestive enzymes break down large food molecules into small soluble ones that can be absorbed.

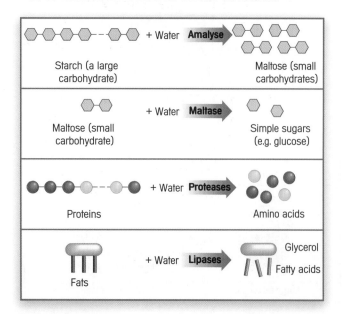

### Quick Test

1. Explain why food must be chemically digested.
2. Give three examples of mechanical digestion and explain why mechanical digestion is important.
3. What is the difference between ingestion and egestion?
4. Describe the structure of a villus and explain how it is adapted to absorb digested nutrients.

**Key Words** — Villi

# Nutrition in Humans

## Measuring the Energy in a Food Sample

When food burns it releases the energy that it contains as heat. You can use the energy to heat up a known volume of water. You can then use the increase in temperature of the water to work out how much energy has been released.

1. Take a sample of **dry food**, (such as a nut) and **weigh it**.
2. Fix the nut on the end of a mounted needle.
3. **Put 25cm³ water into a test tube** and clamp the tube. This much water weighs 25g.
4. **Record the temperature** of the water.
5. Hold the nut in a Bunsen flame until it lights, then immediately hold the burning nut under the test tube of water.
6. If the flame goes out, relight the nut and hold it under the test tube.
7. **Record the temperature** of the water as soon as the nut has finished burning.
8. Record **the increase in temperature** of the water that has been **caused by the energy released by the nut**.

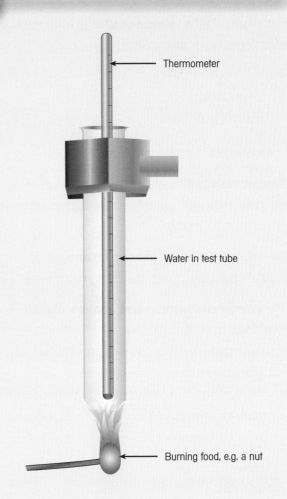

### Calculating the Energy Content of Food

You can use the following equation to calculate the energy that was released by the nut:

$$\text{Energy in food (J)} = 4.2 \times \text{Mass of water (g)} \times \text{Increase in water temperature (°C)}$$

- 1cm³ of water **has a mass** of 1g.
- 4.2 joules of energy can raise the temperature of 1g of water by 1°C.

Therefore:
- 4.2 × 25 joules of energy can raise the temperature of 25g of water by 1°C
- 4.2 × 25 × (increase in temperature) joules of energy can raise the temperature of 25g of water by the increase in temperature that was recorded.

This tells you how much energy was released by the burning nut. Suppose that 25g of water increases in temperature by 10°C. This would mean that 25 × 4.2 × 10 joules of energy had heated the water up.

In this experiment a lot of energy is lost to the surroundings. You can reduce heat loss by insulating the tube and putting a loose foil cap over the top of it.

If you need to compare the energy in different foods, you need to work out how much energy is released per gram of the food:

$$\text{Energy per gram of food (J/g)} = \frac{\text{Energy in food (J)}}{\text{Mass of food (g)}}$$

# Respiration

## Types of Respiration

Respiration is the **release of energy** from food chemicals in all living cells.

The two types of respiration are:
- aerobic respiration
- anaerobic respiration.

### Aerobic Respiration

**Aerobic respiration** releases energy inside living cells by breaking down glucose and combining the products with **oxygen**.

Aerobic respiration needs oxygen and occurs in animal cells, plant cells and in many microbial cells.

The **energy released** is **used in many chemical reactions**, including:
- **movement**, e.g. the contraction of muscles when running
- **the synthesis (making) of large molecules from smaller ones**, e.g. chlorophyll from glucose
- **active transport** of some chemical molecules across a cell membrane.

The equation for this is:

glucose + oxygen → carbon dioxide + water + energy released

$$C_6H_{12}O_6 + 6O_2 \rightarrow 6CO_2 + 6H_2O + \text{energy released}$$

### Anaerobic Respiration

**Anaerobic respiration** releases energy inside the cytoplasm of living cells by breaking down glucose molecules **without the use of oxygen**.

Anaerobic respiration occurs in conditions of very low oxygen, or where no oxygen is present. For example:
- when plant root cells are in waterlogged soil, e.g. rice plants
- in human muscle cells during vigorous exercise, e.g. a 100m sprint
- in bacterial cells inside a puncture wound.

The equation for anaerobic respiration in plant cells and in some microbial cells, e.g. yeast, is:

glucose → carbon dioxide + ethanol + energy released

The equation for anaerobic respiration in animal cells and in some bacteria is:

glucose → lactic acid + energy released

**Aerobic respiration is much more efficient and releases much more energy per glucose molecule (19 times more) than anaerobic respiration.**

**Key Words**: Aerobic respiration • Anaerobic respiration

# Respiration

## Respiration Experiments

### Carbon Dioxide Produced During Respiration

You can show that carbon dioxide is released from living organisms by using an indicator. **Hydrogen-carbonate indicator** is usually red. When it mixes **with carbon dioxide the hydrogen-carbonate indicator turns yellow.**

1. Put a small amount of red hydrogen-carbonate indicator into a test tube.
2. Lower a gauze platform into the tube with tweezers.
3. Put some living organisms (e.g. small insects like woodlice or soaked germinating seeds) onto the platform. Seal the tube.

**Respiration from the living organisms will release carbon dioxide** and the indicator will turn yellow.

You will need a control tube as well. The control should either have dead organisms in it or none at all. You can then be sure that the **carbon dioxide is coming from the living processes** in the organisms.

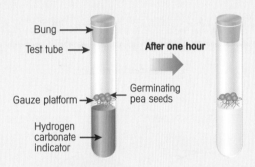

### Temperature on the Rate of Respiration

The method above can be used to investigate the effect of temperature on the rate of respiration. Set up the tube in a water bath at a controlled temperature.

Time how long it takes for the hydrogen-carbonate indicator to turn yellow and then repeat the experiment at different temperatures.

### Energy Released During Respiration

When living organisms respire, they release energy that can be detected as heat. In this experiment, germinating peas produce heat energy, whereas boiled (dead) peas do not.

1. Soak peas in water and mild disinfectant to start germination and kill microorganisms on the peas.
2. Boil half the peas.
3. Set up two thermos flasks, as shown and record their temperatures for a week.

In the test flask the **peas respire, giving off energy as heat**. The flask keeps the heat in, so **the temperature rises**. In the **control flask, no respiration can occur** because the peas are dead. The temperature stays the same, unless the air inside the flask is affected by temperature changes in the surroundings. The difference in temperature between the flasks is due to heat produced during respiration.

## Quick Test

1. Name three ways in which cells use energy.
2. What is respiration?
3. How is aerobic respiration different from anaerobic respiration?
4. In an experiment, living peas are kept in a thermos flask. Why does the temperature of the flask increase?

**Key Words**  Control

# Gas Exchange in Plants

## Understanding Gas Exchange in Plants

During **respiration** plants use up oxygen and produce carbon dioxide. During **photosynthesis**, plants use up carbon dioxide and produce oxygen.

Carbon dioxide ($CO_2$) and oxygen ($O_2$) move in and out of plant leaves by diffusion through the **stomata** (tiny pores). Inside the leaf, carbon dioxide and oxygen are exchanged by diffusion between the air and the plant cells.

## Diffusion of Carbon Dioxide and Oxygen in Plants

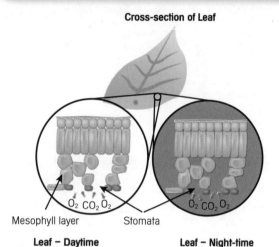

**Cross-section of Leaf**

Mesophyll layer   Stomata

**Leaf – Daytime**
In the light, the plant photosynthesises and **uses carbon dioxide faster** than it is made during respiration. The plant also **makes oxygen from photosynthesis faster** than it uses it in respiration.

**Leaf – Night-time**
When there isn't enough light for photosynthesis, the carbon dioxide from respiration is released and oxygen needed is taken up.

The **stomata** on the underside of leaves are specially adapted to:
- **open** – to help increase the rate of diffusion of carbon dioxide and oxygen
- **close** – to prevent excessive water loss in drought conditions.

(P2) **Plants respire all the time, during the day and the night. Plants also photosynthesise, but only when there is enough light.** During **photosynthesis** a plant uses up carbon dioxide and produces oxygen. If there is a high intensity of light the plant photosynthesises faster than it respires. Then there is a **net uptake of carbon dioxide and a net loss of oxygen from the plant.**

## (P2) Using Hydrogen-carbonate Indicator

You can use hydrogen-carbonate indicator to investigate the **effects of different light intensity** on **gas exchange** in leaves.

Hydrogen-carbonate indicator is red when mixed with carbon dioxide at the concentration normally found in air. It becomes different colours when the concentration of carbon dioxide changes.

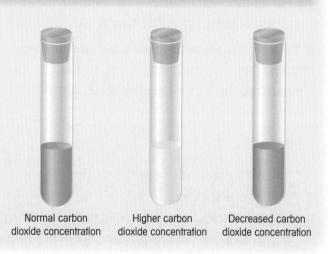

Normal carbon dioxide concentration   Higher carbon dioxide concentration   Decreased carbon dioxide concentration

**Key Words**   Stomata

# Gas Exchange in Plants

## Investigating Gas Exchange in Plants

The following experiment shows how light affects gas exchange in plants.

1. Set up four test tubes containing red hydrogen-carbonate indicator. **Tube A is the control**.
2. Tie thread to the leaf stalks of three fresh leaves and suspend the leaves in the three non-control test tubes so the leaves are hanging above the indicator solution. You can trap the thread in the bung.
3. Wrap **tube B** completely in foil so **no light** can get to the leaf.
4. Wrap **tube C** in a thin layer of gauze so **a little light** can reach the leaf.
5. Leave **tubes D and A** unwrapped so these get **plenty of light**.
6. Put all the tubes in bright light for an hour, then unwrap them and record the colours straight away.

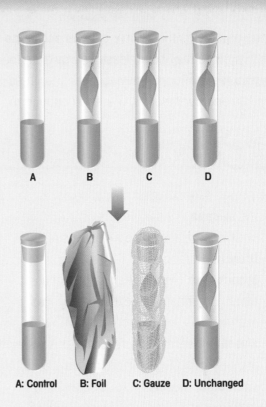

A: Control  B: Foil  C: Gauze  D: Unchanged

### The Results

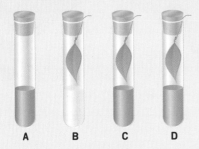

**In tube A** the indicator is still mixing with air in the tube at normal carbon dioxide concentration, so it stays red. **Tube A is the control**.

**In tube B** the leaf **cannot photosynthesise**, but it is **still respiring**. So the leaf has given off carbon dioxide, which **increases the carbon dioxide concentration** in the air. This makes the indicator go yellow.

**In tube C** the leaf **is respiring but photosynthesis is slow**. This is because only a little light can reach the leaf through the gauze. The leaf is using up carbon dioxide for photosynthesis **at about the same rate** that it is producing carbon dioxide from respiration. There is **not much change of carbon dioxide concentration** in the tube, so the indicator stays red.

**In tube D** the **leaf is respiring but lots of photosynthesis** is taking place. **So more carbon dioxide is being used up than is being produced** by respiration. The concentration of **carbon dioxide in the air in the tube decreases** and this turns the indicator purple.

## Quick Test

1. When does a plant:
   a) respire?  b) photosynthesise?
2. What are stomata and why are they important?
3. Why are air spaces in leaves important?
4. Explain why plants release oxygen during the day and carbon dioxide at night.

# Gas Exchange in Humans

## Structures for Gas Exchange

Living organisms must carry out **gas exchange** to get oxygen so they can release energy from food by **aerobic respiration**. Some organisms, such as amoeba and earthworms, are small enough to obtain oxygen by **diffusion** through their moist permeable skin. But bigger, more complex organisms need specialised structures like lungs or gills to obtain oxygen.

## The Human Thorax

The human **thorax** (chest cavity) contains:
- the **trachea** – a flexible tube, surrounded by rings of cartilage to stop it collapsing
- **bronchi** – branches of the trachea
- **bronchioles** – branches of a bronchus
- **lungs** – to inhale and exhale air for gas exchange
- **alveoli (air sacs)** – microscopic air sacs at the ends of bronchioles, where gases are exchanged
- **intercostal muscles** – to raise and lower the ribs
- **pleural membranes** – to protect and lubricate the surface of the lung
- the **diaphragm** – a muscular 'sheet' between the thorax and abdomen.

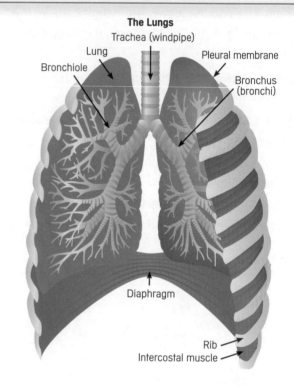

The Lungs

## The Alveoli

Carbon dioxide diffuses from the blood into the alveolus, and oxygen diffuses from the alveolus into the blood. This is called **gas exchange**.

The lungs are adapted for gas exchange because:
- there are millions of microscopic alveoli, which provide an enormous surface area for gas exchange
- the shape of each alveolus increases the surface area for gas exchange
- each alveolus has a moist, thin, permeable surface that gases diffuse through easily
- each alveolus has an excellent blood supply and is kept supplied with fresh air, so gases are carried away from the gas exchange surface easily.

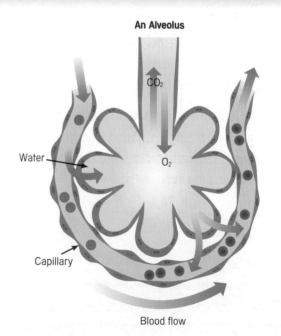

An Alveolus

**Key Words**: Trachea • Bronchi • Bronchioles • Alveoli

# Gas Exchange in Humans

## Breathing

During **breathing**, the volume and pressure of the chest cavity are changed by:
- the intercostal muscles
- the diaphragm.

The diagrams below show **inspiration** and **expiration**.

The movement of air in and out of the lungs is called **ventilation**. Ventilation is brought about by breathing.

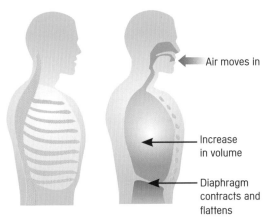

**Inspiration**
Intercostal muscles contract; ribcage raised

- Air moves in
- Increase in volume
- Diaphragm contracts and flattens

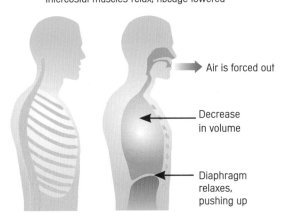

**Expiration**
Intercostal muscles relax; ribcage lowered

- Air is forced out
- Decrease in volume
- Diaphragm relaxes, pushing up

## Protection against Disease

The **respiratory system** has defences to protect itself from disease.

The **trachea** and **bronchi**:
- produce **mucus** to trap dust and microorganisms
- are lined with millions of **cilia** (ciliated cells), which move the mucus (with dust and microorganisms) from the lungs into the throat, where it's swallowed.

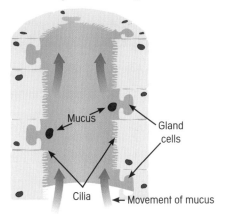

**A Breathing Tube in the Lungs**

- Mucus
- Gland cells
- Cilia
- Movement of mucus

## Quick Test

1. Name the structures air passes through as it is breathed in.
2. Explain why cartilage rings are needed around the trachea.
3. Describe what happens to oxygen and carbon dioxide at the gas exchange surface of the alveolus.
4. Name the muscles used in ventilating the lungs.
5. Describe how air is: **a)** moved into the lungs **b)** pushed out of the lungs.
6. Describe how particles of dust are removed from the lungs.

**Key Words**  Inspiration • Expiration • Ventilation • Respiratory system

# Gas Exchange in Humans

## Smoking and Lung Disease

**Smoking** can lead to several diseases, including **cancer** of the mouth, throat, oesophagus and lungs, heart disease, **emphysema** and bronchitis.

Smoking damages the **cilia** (ciliated epithelial cells), which line the airways (**trachea, bronchi** and **bronchioles**). This prevents the cilia from being able to remove mucus, tar and dirt from the lungs, which leads to a 'smoker's cough' as the body tries to cough up mucus. Excess coughing can damage the alveoli and cause emphysema.

The smoking machine experiment shows that cigarettes contain **tar**. Cigarettes also contain **nicotine**, which is very addictive, and produce **carbon monoxide** and particulates when they are burned. The carbon monoxide produced by a burning cigarette is dangerous because the blood picks it up instead of picking up oxygen. This means the blood is carrying much less oxygen, which makes smokers breathless.

Tar contains chemicals that are **irritants** and **carcinogens** (cancer-causing chemicals). **Particulates** in cigarette smoke accumulate in living tissue (e.g. lungs) and can cause cancer.

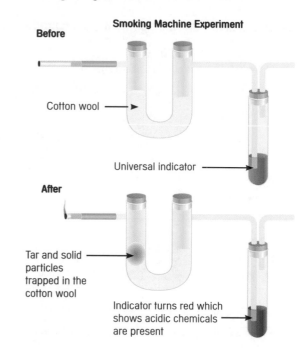

## Smoking and Diseases of the Circulatory System

Smoking increases the risk of getting **cardiovascular disease** – this **includes heart attacks and strokes**:
- Substances in smoke damage the lining of arteries and increase the risk that the arteries will become narrowed by fatty material (**atheroma**). This can cause angina, heart attacks and strokes.
- **Coronary heart disease** is when **atheroma narrows the coronary arteries**. The coronary arteries then cannot supply as much oxygenated blood to the muscle of the heart.
- **Nicotine** in smoke makes your heart beat faster and **increases your blood pressure**. Your heart then has to work harder.
- **Smoking** increases the risk **that blood will clot** – another cause of heart attacks and strokes.

- **Carbon monoxide** in **smoke lowers the amount of oxygen blood can carry**. The heart then has to pump harder to get the oxygen needed around the body. In a pregnant smoker, the **developing foetus gets less oxygen and so grows more slowly**.

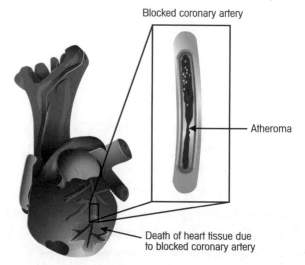

**Key Words**  Emphysema • Carcinogen • Cardiovascular disease • Atheroma

# Gas Exchange in Humans

## Exercise and Muscle Cells

To get the energy needed for exercise, muscle cells **need lots of oxygen and glucose for respiration**.

The energy released causes the muscle to contract and move the body.

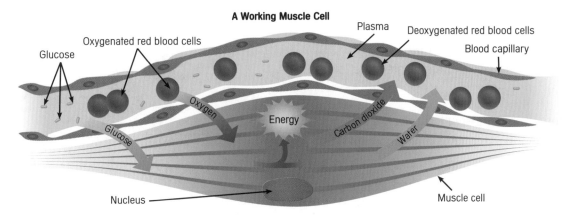

A Working Muscle Cell

## Exercise, Heart Rate and Breathing Rate

When you exercise, your **muscles contract more** frequently and more strongly. They need to release more energy in order to do this, so **they respire faster**.

For respiration to go faster, oxygen and glucose will need to be delivered faster to the cells, and more carbon dioxide will have to be removed from the body.

Blood must flow faster in order to deliver oxygen to the cells faster and remove more carbon dioxide from them. **The heart rate therefore speeds up** to push blood round the body more quickly.

You can investigate the effect of exercise on **breathing rate** in a simple experiment:

1. Sit still for five minutes and then count your breaths for one minute.
2. Now do some exercise for four minutes. The exercise could be standing on and off a step, skipping or jogging on the spot, etc.
3. Immediately after you have finished exercising for four minutes, count how many breaths you take in one minute.
4. Record the change in number of breaths per minute before and after exercise.
5. Repeat the experiment a few times with different people.

Variables that you should keep the same include the time spent exercising and the type of exercise you do.

You should find that the number of breaths per minute increases during exercise. If you count the breaths per minute every couple of minutes after finishing exercise you will find that the breathing rate stays higher than normal for several minutes after the exercise stops.

During exercise, **extra carbon dioxide is produced** by muscle cells. The cells **also need more oxygen**. The increase in breathing rate removes the extra carbon dioxide from the body faster. It also brings fresh air, rich in oxygen, into the alveoli more often. So **oxygen is absorbed into the blood faster**.

### Quick Test

1. List four diseases caused by smoking.
2. List three harmful substances found in tobacco smoke.
3. Explain how carbon monoxide causes breathlessness.
4. Explain why it is necessary for the heart rate to increase during exercise.

# Transport in Plants

## The Need for Transport Systems

Organisms made of one or only a few cells do not need a transport system. This is because **their surface area is very large compared to their volume** and so the substances they need can reach all their cells **by diffusion through their surfaces**. They can also excrete waste materials by diffusion efficiently.

In larger organisms, most of the cells are a long way away from the surrounding environment. **Diffusion is too slow** to bring the materials from the environment to all the cells. **Transport systems bring the materials that cells need very close to every cell**. They also carry waste products away to the surfaces of the organism to be lost to the environment.

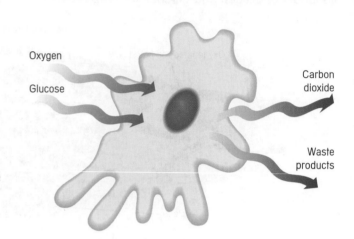

## Understanding Transport Systems in Plants

The **xylem** and **phloem** form a continuous system of tubes, from roots to leaves, called **vascular bundles**:
- **Xylem** transports water and soluble mineral salts from the roots to the leaves.
- **Xylem vessels** are made from dead plant cells. They have a hollow **lumen**. The cellulose cell walls are thickened with a waterproof substance.
- **Phloem** allows the movement of food substances (sucrose and amino acids) around the plant (translocation), up and down stems to growing tissues and storage tissues.
- **Phloem** cells are long columns of living cells.

**Root hairs** have an enormous surface area for absorbing water and so increase the plant's ability to take up water.

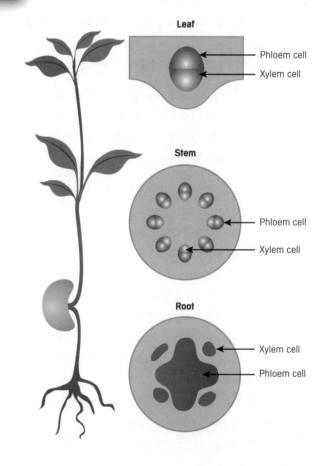

Key Words: Translocation • Root hair

# Transport in Plants

## Absorbing Water

Water gets into the root hair cells of plants by **osmosis**. The concentration of dissolved substance inside the root hair cells is greater than the concentration of substances in the soil. So water diffuses **down the concentration gradient** into the cell. Water then passes across the root cells by osmosis until it reaches the xylem.

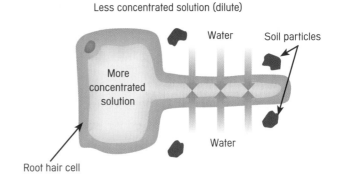

## Transpiration

**Transpiration** is the loss of water vapour from a leaf, by evaporation from cell surfaces and diffusion through the stomata.

It causes water to be moved up xylem vessels and provides plants with water for cooling, photosynthesis and support. The water carries minerals from the root to the leaves.

The transpiration stream is powered by the evaporation of water from the leaf:

1. Water evaporates from the internal leaf cells and diffuses out of the leaf through the stomata.
2. Water passes by osmosis from the xylem vessels to leaf cells, which pull the thread of water in that vessel upwards by a very small amount.
3. Water enters the xylem from root tissues, to replace water which has moved upwards.
4. Water enters root hair cells by osmosis and replaces water which has entered the xylem.

The rate of **transpiration** can be affected by:
- **light intensity** – higher light intensity increases the rate of photosynthesis and transpiration
- **air movement (wind)** – as the movement of the air increases, transpiration increases
- **temperature** – heat increases the rate of photosynthesis and transpiration
- **humidity** – low humidity increases the rate of transpiration.

A leafy shoot's rate of transpiration can be measured using a **mass potometer**:

1. The plant's roots are submerged in a sealed bag of water and placed in a beaker.
2. The beaker is placed on a digital balance.
3. Readings are then taken to see how much water is lost by the plant during transpiration.
4. The conditions (e.g. light, temperature) can be changed to see how this affects water loss.

## Quick Test

1. What is transpiration?
2. Name two substances transported in the: a) xylem b) phloem.
3. Name three environmental factors that speed up the rate of transpiration.
4. Explain how water enters root hair cells from the soil.
5. Describe how you could measure the rate of transpiration.

**Key Words** — Transpiration • Potometer

# Transport in Plants

## Measuring Transpiration

You can use a **potometer** to measure how fast **transpiration** occurs.

### Setting up a Potometer

To set up a potometer, follow these instructions:

1. Cut a leafy shoot, at a slant, **under water so that air doesn't get into the xylem**.
2. Fill a narrow perspex tube with water and fit the shoot into it.
3. Push a graduated pipette, full of water, into the other end of the tube.
4. Clamp the tube into a U-shape.
5. Make sure all the **joins are sealed** so air cannot get in.
6. As the leaves transpire, water moves into the shoot pulling air into the pipette. Start a stop clock and record the **distance the air moves** along the pipette in **a set time**, e.g. 20 minutes.
7. Calculate and record the **distance moved per minute**.
8. Unclamp the pipette and put the open end into a beaker of water. Then squeeze the tube a few times to push the air bubble out and refill the pipette with water.
9. Re-clamp the pipette and repeat the experiment.

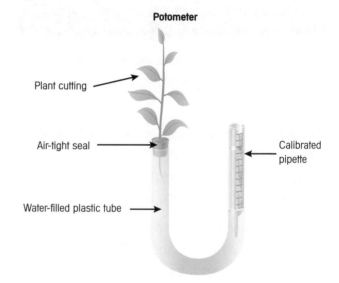

Potometer

### An Alternative Potometer

Here is a different kind of potometer. You can first let a bubble of air enter the horizontal glass tube and then you can record how far the bubble travels per minute. To repeat the reading, you will have to push the bubble back out of the tube. How could you do that?

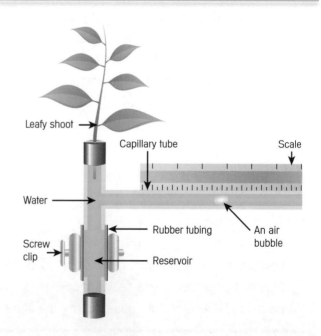

# Transport in Plants

## Investigating Factors Affecting Transpiration

You can use potometers to compare the rates of transpiration in different environmental conditions:

1. Place each potometer in a different environment, e.g. mist, wind and bright light. Keep one potometer in normal room conditions as the control.

A Control room conditions   B Mist   C Wind

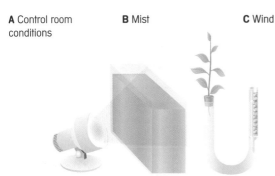

D Bright light

2. Measure water loss in each potometer every 3 minutes for 30 minutes.

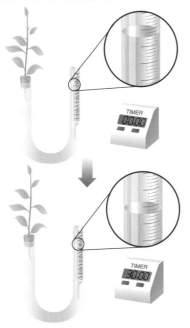

### Results and Explanation

**Increasing transpiration** may be due to the following factors:

- **High light intensity** – causes the stomata to open.
- **High temperatures** – increase the rate of evaporation of water from cells inside the leaf which **increases the concentration gradient** for water vapour between the leaf air spaces and the atmosphere. The steeper concentration gradient makes water vapour **diffuse more quickly out** through the stomata.
- **Wind blowing water vapour away** from the air around the stomata. There is then a much higher concentration of water vapour in the air spaces of the leaf than in the surrounding air. The **concentration gradient for water vapour remains steep**, so water vapour **diffuses quickly** out of the leaf.

**Decreasing transpiration** may be due to **high humidity**:

- There is a high concentration of water vapour in humid air. Humid air outside the stomata reduces the concentration gradient between the leaf air spaces and the surrounding air. Water vapour then only **diffuses slowly out of the leaf**.

### Quick Test

1. What is a potometer used for?
2. Explain why high light intensity increases the rate of transpiration.
3. Explain how and why an increase in temperature affects the rate of transpiration.
4. Explain how high humidity affects the rate of transpiration.
5. Under what environmental conditions would you expect a plant to be most likely to wilt?

# Transport in Humans

## Blood

Blood has **four** components – **platelets** (bits of broken cells), **plasma, white blood cells** and **red blood cells**.

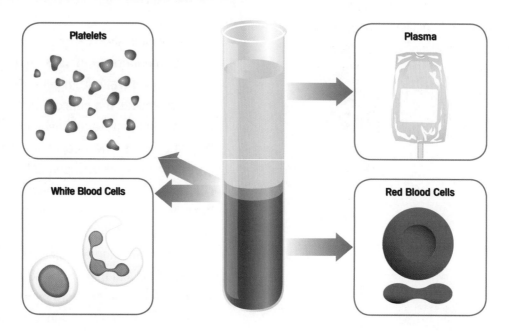

**Plasma is a straw-coloured liquid** which transports dissolved carbon dioxide, digested food, urea, hormones and heat energy around the body.

**White blood cells** protect the body against disease. Some have a flexible shape which allows them to engulf disease-causing microorganisms.

**Red blood cells** transport oxygen from the lungs to the tissues. The shape, structure and presence of haemoglobin make red blood cells well adapted for this function:
- They are small and flexible, so they can pass through narrow blood vessels.
- They don't have a nucleus, so they can be packed with **haemoglobin**.
- The small size and biconcave shape of red blood cells gives them a **large surface area to volume ratio** for absorbing oxygen. When the cells reach the lungs, oxygen diffuses from the lungs into the blood.

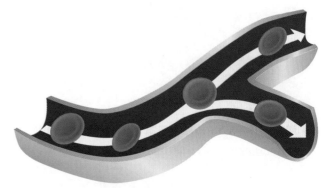

The haemoglobin molecules in the red blood cells binds easily with the oxygen to form **oxyhaemoglobin**.

**haemoglobin + oxygen = oxyhaemoglobin**

The blood is then pumped around the body to the tissues, where the reverse reaction takes place. Oxygen is released which diffuses into cells.

(P2) **Platelets** clump together when a blood vessel becomes damaged in order to produce a **clot**. Clotting prevents blood loss and stops microorganisms entering the body.

Key Words: Haemoglobin • Oxyhaemoglobin

# Transport in Humans

## The Circulatory System

Blood moves around the body in the blood vessels – **arteries**, **veins** and **capillaries**.

The heart pumps blood through the blood vessels:
- The **right-hand side** of the heart pumps blood which is low in oxygen **to the lungs and back**.
- The **left-hand side** of the heart pumps blood which is rich in oxygen **to the rest of the body and back**.
- Blood pumped into the arteries is under much higher pressure than the blood in the veins.

Mammals have a **double circulatory system**, i.e. it consists of two loops. The advantage of this is that blood is pumped to the body at a higher pressure than it is pumped to the lungs. This provides a much greater rate of flow to the body tissues.

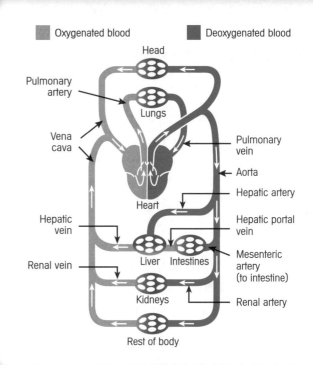

## Blood Vessels

The main blood vessels are **arteries**, **veins** and **capillaries**. Their structure is related to their function.

| **Arteries** carry blood **away** from the heart **towards** the organs. Substances from the blood can't pass through artery walls. An artery has a thick, elastic, muscular wall to cope with the high pressure in the vessel. | **Veins** carry blood **from** the organs **back** to the heart. Substances can't pass through the veins' walls. A vein has a thinner wall than an artery and has less elastic muscular fibre due to the lower pressure in the vessel. Veins have pocket valves along their length to keep blood flowing in the right direction. | **Capillaries** are very narrow vessels that carry blood between arteries and veins, and through the organs. They have walls made of a single layer of cells. It's here that substances are exchanged between blood and body cells. |
|---|---|---|
| 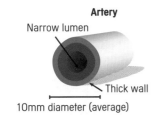 Artery – Narrow lumen, Thick wall, 10mm diameter (average) | 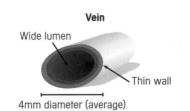 Vein – Wide lumen, Thin wall, 4mm diameter (average) | 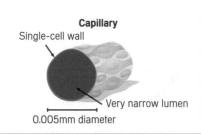 Capillary – Single-cell wall, Very narrow lumen, 0.005mm diameter |

## Quick Test

1. List the components of the blood.
2. Give three adaptations of red blood cells that help them to perform their function.
3. Describe and explain the differences between the structures of arteries and veins.

**Key Words**: Artery • Vein • Capillary

# Transport in Humans

## The Heart

The **heart** consists of powerful muscles that pump blood around the body. It needs glucose and oxygen for respiration because it never gets tired or needs rest, so it has high energy requirements.
- The **coronary artery** supplies the heart itself with glucose and oxygen.
- The **pulmonary vein** carries **oxygenated blood** from the lungs to the heart.
- The **aorta** carries oxygenated blood from the heart to the rest of the body.
- The **vena cava** carries **deoxygenated blood** from the parts of the body back to the heart.
- The **pulmonary artery** carries deoxygenated blood from the heart to the lungs.

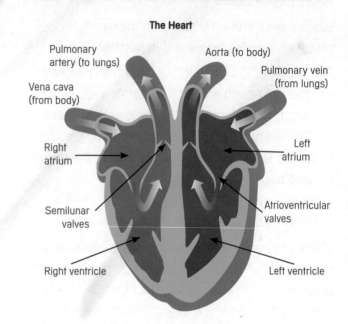

## The Cardiac Cycle

The cardiac cycle is the sequence of events that occurs when the heart beats.

During each heart beat, the following takes place:
1. The heart relaxes and blood enters both **atria** from **veins**.
2. The atria contract to push blood into the **ventricles** through the atrioventricular valves.
3. The ventricles contract, pushing blood into the arteries. The semilunar valves open to allow this whilst the atrioventricular valves close.

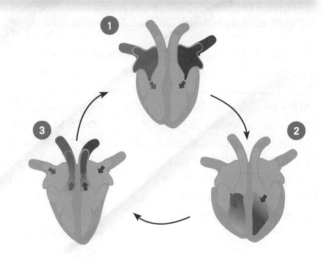

## Controlling Heart Rate

The heart beat is controlled by groups of cells called the **pacemaker** (or sino-atrial node).

The **sino-atrial node (SAN)** produces impulses that spread across the atria to make them contract.

The **atrioventricular node (AVN)** relays impulses that spread over the ventricles to make them contract.

Nerves connecting the heart to the brain can increase or decrease the pace of the SAN in order to regulate the heart beat. During exercise, muscles demand more energy so the pacemaker fires more frequently. This makes the heart rate speed up to supply **oxygen** and **glucose** to respiring muscles more efficiently.

Adrenaline is also secreted during exercise. Adrenaline acts on the pacemaker making it fire more frequently.

**Key Words**: Oxygenated • Deoxygenated • Pacemaker

# Transport in Humans

## Microorganisms and Infection

Infections are caused by microorganisms damaging body cells or producing poisons (toxins) that harm cells. Infections can be treated with drugs called **antimicrobials** (e.g. **antibiotics**).

Many antimicrobials kill the microorganisms but some just block or slow down their action. This is called **inhibition**.

Microorganisms that cause infections are called **pathogens** and include:
- **bacteria**, which cause bubonic plague, tuberculosis (TB) and cystitis – treated by antibiotics
- **fungi**, which cause athlete's foot, thrush and ringworm – treated by anti-fungal medicine and antibiotics
- **viruses**, which cause Asian bird flu, common cold, HIV, measles and smallpox – very difficult to treat; antibiotics don't work on viruses.

Microorganisms can be found on any surface, in food and drink, in water and in the air we breathe.

The body provides ideal conditions for microorganisms to grow – it's **warm**, with plenty of **nutrients** and **moisture**. Once in your body, harmful microorganisms reproduce very rapidly – some populations can double as fast as every 20 minutes.

Symptoms of an illness only show when there's a significant amount of infection. The symptoms are caused by microorganisms damaging infected cells, for example bursting the cells or producing harmful toxins.

## The Immune Response

If microorganisms get into your body, your **immune system** is activated.

Two types of **white blood cell** are important in this response – **phagocytes** and **lymphocytes**.

1. One type of white blood cell (phagocyte) is activated when any microorganism gets into the body.
2. Another type of white blood cell (lymphocytes) makes **antibodies**.

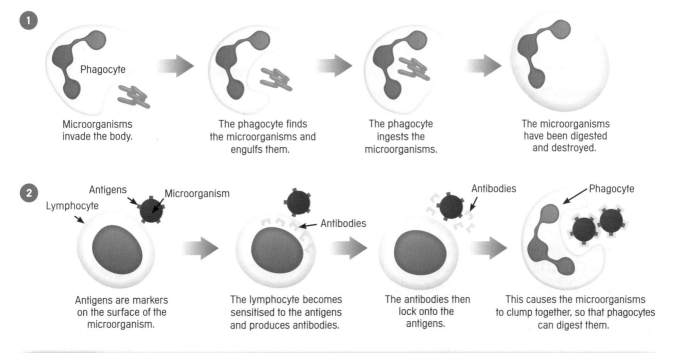

**Key Words**  Antibiotic • Bacteria • Immune system • Antibody

# Transport in Humans

## Specialisation of Antibodies

Different microorganisms cause different diseases.

Microorganisms have **unique markers**, called **antigens**, on their surface. White blood cells produce antibodies specific to the marker they need to attack. After infection, some white cells act as memory cells as they 'remember' the antigens and are able to produce antibodies quicker if the body is re-infected. This is **natural immunity** and it protects that particular individual in the future.

Example – Antibodies to Fight TB will not Fight Cholera

##  Vaccination

A **vaccination** helps the body to develop **immunity** and produce **specific antibodies** and **memory cells** so that microorganisms can be destroyed before they **cause infection**. Vaccines contain antigens that are the same as those found in the microorganism that causes a disease.

When the vaccine is injected it stimulates an immune response to these antigens, but does not cause the illness that the microorganism would do. Memory cells are produced in response to the antigen. The memory cells cause a much faster immune response if the vaccinated person is later infected by microorganisms with the same antigens as the vaccine. So the microorganisms are destroyed before they can cause illness.

Vaccinations are never completely safe and can produce **side effects**, most of which are minor, like rashes. **Extreme side effects** are **rare**, but the vaccination carries less risk than the disease. However, because of genetic variation, some people may be affected more than others. But the benefits to the majority are usually considered greater than the risks, so this course of action is often recommended by the government.

Some vaccines have to be redeveloped regularly because microorganisms **mutate** (randomly genetically change) to produce **new varieties (strains)**. For example, **flu vaccinations** are **renewed every year** because new strains appear.

1. A weakened/dead strain of the microorganism is injected. Antigens on the modified microorganism's surface cause the lymphocytes to produce specific antibodies.

2. The lymphocytes that are capable of quickly producing the specific antibody remain in the bloodstream.

## Quick Test

1. Explain why microorganisms reproduce very rapidly once inside the body.
2. Name two types of white blood cell that are important in the immune response.
3. Describe how antibodies help to destroy pathogens.
4. Explain how vaccinations prevent illness.

**Key Words** — Antigen • Natural immunity

# Excretion

## Organs of Excretion

**Excretion** is the removal, from an organism, of waste substances produced by chemical reactions inside its cells (**metabolism**).

Leaves are excretion organs for a plant. Respiration and photosynthesis are **metabolic processes** which take place in plant cells.

The skin, lungs and kidneys are organs used for excretion.

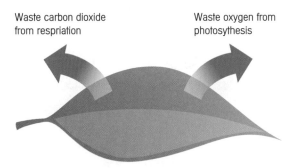

Waste carbon dioxide from respriation

Waste oxygen from photosythesis

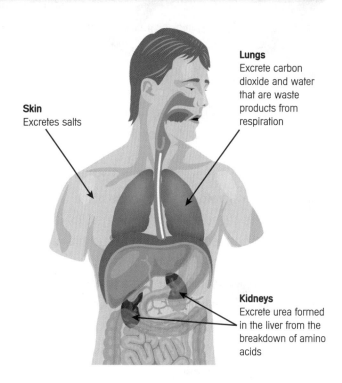

**Skin**
Excretes salts

**Lungs**
Excrete carbon dioxide and water that are waste products from respiration

**Kidneys**
Excrete urea formed in the liver from the breakdown of amino acids

## The Kidneys

Urea is a waste product made when the liver breaks down excess amino acids. **Urea is excreted from the blood by the kidneys.**

The kidneys also **regulate** how **much water and salt** is kept in the blood, a process called **osmoregulation**.

The kidneys filter most of the small molecules out of the blood at high pressure. They then reabsorb useful substances in the amounts needed by the body. **Glucose, amino acids, some salts and some water are reabsorbed. Urea, excess salts and excess water** are not reabsorbed but **are lost as urine**.

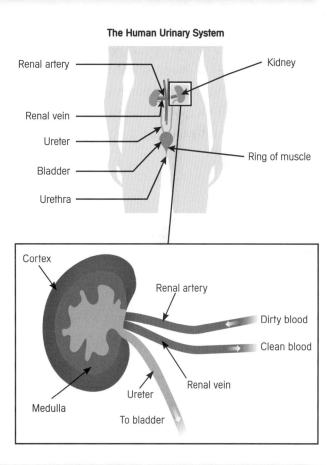

The Human Urinary System

Renal artery, Kidney, Renal vein, Ureter, Bladder, Ring of muscle, Urethra

Cortex, Renal artery, Dirty blood, Clean blood, Renal vein, Ureter, Medulla, To bladder

**Key Words** — Metabolism • Osmoregulation

# Excretion

## How Kidneys Work

The kidneys contain millions of tiny filtering units called **nephrons**. Three stages take place:

1. **Ultrafiltration** – the blood from the renal artery is forced into the glomerulus under high pressure. Most of the water is forced out of the glomerulus and into the Bowman's capsule, including all the small molecules like urea, salts and glucose. The liquid filtered into the Bowman's capsule is called **glomerular filtrate**. The cup shape of the Bowman's capsule allows the glomerulus (a big ball of blood vessels) to sit inside and allows high pressure to be maintained.

2. **Selective reabsorption** – in the proximal convoluted tubule useful substances, such as glucose, are reabsorbed into the blood, which runs very close by. The coiled-up tubule is long and folded to allow time for the useful substances to pass back into the blood.

3. **Salt and water regulation** – the collecting duct is where water is reabsorbed into the blood. The hairpin-shaped loop of Henlé and the collecting duct extend from the cortex to the medulla, allowing plenty of time for reabsorption of water and ions. Complex movements of ions and water across the loop result in the production of concentrated urine.

Urine contains salt, water and urea.

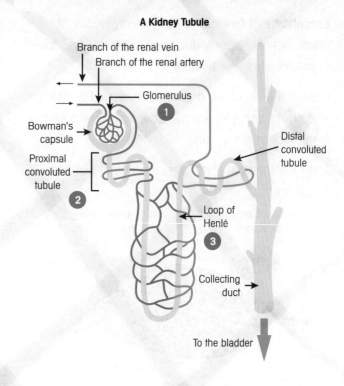

A Kidney Tubule

## Regulating Water in the Body

Water is **input** (gained) from:
- food and drinks
- respiration.

Water is **output** (lost) through:
- sweating
- breathing
- excretion of faeces and urine.

Your body has to **balance** these different inputs and outputs to ensure that there's enough water inside cells for cell activity to take place.

The kidneys play a vital role in balancing water, waste and other chemicals in the blood.

**Key Words** — Ultrafiltration

# Excretion

## Controlling Reabsorption of Water

The concentration of urine is controlled by a hormone called **anti-diuretic hormone (ADH)**, which is released into your blood via the **pituitary gland**.

Controlling water balance is an example of **negative feedback**, where one system is the reverse to another in order to maintain a steady state.

### When There Is Too Much Water in the Blood

When your blood water level becomes **too high** (i.e. there's too much water), the following happens:

1. **Receptors** in your **hypothalamus** detect a **decrease** in salt concentration. No stimulus is sent to the pituitary gland.
2. **Less** ADH is secreted into the blood.
3. The collecting ducts of your kidneys become **less permeable**, so less water is reabsorbed.
4. Your bladder fills with a **large quantity of dilute urine**.

### When There Is Not Enough Water in the Blood

If your blood water level becomes **too low** (i.e. not enough water), the opposite happens:

1. Receptors in your **hypothalamus** detect an **increase** in salt concentration. A stimulus is sent to the pituitary gland. **Thirst** is stimulated to encourage drinking.
2. **More** ADH is secreted into the blood.
3. The collecting ducts of your kidneys become **more permeable**, so more water is reabsorbed.
4. Your bladder fills with a small quantity of **concentrated urine**.

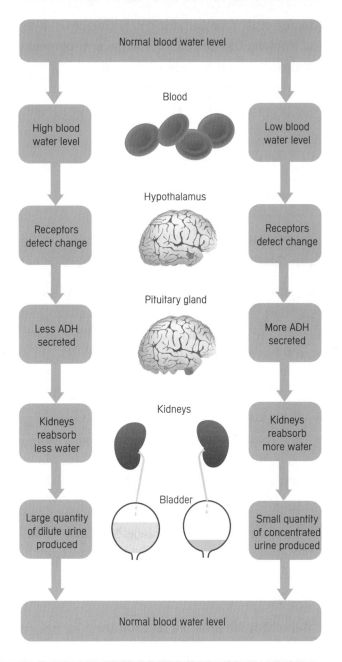

## Quick Test

1. What is excretion?
2. Name two substances excreted from plant leaves and state where these substances come from.
3. Starting from the glomerulus, name the structures that a urea molecule passes through on its way from the blood to the bladder.
4. Glucose is filtered out of the blood in the kidney. Why is glucose **not** found in the urine of healthy people?
5. Name the hormone involved in regulating water loss in the urine.
6. How does an increase in ADH affect the colour, concentration and volume of urine?

**Key Words**  Anti-diuretic hormone (ADH) • Pituitary gland • Hypothalamus

# Coordination and Response in Humans

## Responding to the Environment

All living organisms can respond to changes in the environment. This can be the **internal environment**, within and around the cells of the organism. Or it can be the **external environment** that surrounds the body of the organism. A change in the environment is called a **stimulus**.

Your body has automatic control systems to keep the internal environment constant so that cells can function efficiently. This is called **homeostasis**. It is the maintenance of a constant internal environment.

Your **kidneys regulate the water content** of the internal environment by secreting the hormone ADH. By doing this the concentration of the internal environment is kept constant.

Keeping the temperature of the body constant is another example of homeostasis.

## Temperature Control

**Enzymes** in your body work best at **37°C**, so it's essential that your body remains very close to this temperature. Heat produced through respiration is used to maintain your body temperature.

### If body temperature becomes too high

- Blood vessels in the skin widen (vasodilation) and the blood flows closer to the skin surface so heat can be lost
- Sweat is produced – the evaporation of sweat requires heat energy.

Getting too hot can be very dangerous. If too much water is lost through sweating, the body becomes **dehydrated**. This can lead to **heat stroke** and even **death**.

### If the body temperature falls too low

- Blood vessels in the skin constrict (vasoconstriction) and the blood flow near the skin is reduced.
- Sweating stops
- Muscles start making tiny contractions (shivering) which releases heat.

Getting too cold can be fatal. **Hypothermia** is when the body temperature drops below 35°C. This causes **unconsciousness** and sometimes **death**.

Putting on more clothing and doing exercise can help to raise body temperature.

**Negative feedback** involves the automatic reversal of a change in condition. It occurs frequently in homeostasis. For example, if the temperature falls too low, the brain switches on mechanisms to raise it. If the temperature then becomes too high, the brain switches on mechanisms to lower it.

**Vasodilation – When the Body is too Warm**

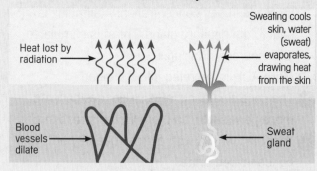

**Vasoconstriction – When the Body is too Cold**

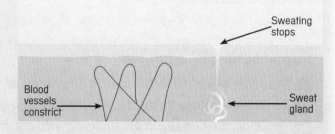

48 | Key Words | Stimulus

# Coordination and Response in Humans

## Hormones

**Hormones** are chemical messages released by **glands** directly into the bloodstream. They travel around the body to their target organs.

Hormones travel much **slower** (at the speed of blood) than a nervous impulse, which is an electrical message relayed directly to the target organs.

Hormones are often released by endocrine glands when **a change in the internal environment** acts as a stimulus. **Receptors detect the stimulus** and hormones are released which cause activity in the target organ. The **target organ is an effector**.

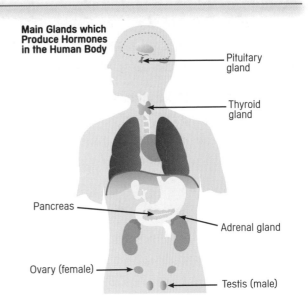

Main Glands which Produce Hormones in the Human Body
- Pituitary gland
- Thyroid gland
- Pancreas
- Adrenal gland
- Ovary (female)
- Testis (male)

## Action of Hormones

| Hormone | Source of Hormone | Role of Hormone | Effect of Hormone on Target Organ |
|---|---|---|---|
| ADH | Anterior pituitary gland (brain) | Regulates the amount of water in the body | Increase in ADH increases the amount of water reabsorbed by the kidneys |
| Adrenaline | Adrenal glands | Increases the heart and breathing rate, the blood sugar level and the rate of respiration | More energy is released in the muscles, so the body is ready for 'action' (fight or flight responses) |
| Insulin | Pancreas | Decreases the level of glucose in the blood | Causes liver cells to change blood glucose into glycogen to be stored |
| Testosterone | Testes | Controls male sexual characteristics | Causes development of male sexual characteristics, e.g. facial hair and development of sperm |
| Progesterone | Ovaries | Maintains pregnancy | Keeps the lining of the uterus intact; regulates menstrual cycle |
| Oestrogen | Ovaries | Main female sex hormone | Regulates menstrual cycle and promotes female sexual characteristics; causes development of uterus lining |

## Quick Test

1. What is homeostasis?
2. Give two examples of conditions that are kept constant in your body.
3. (P2) Describe the changes that take place in your body when your temperature begins to rise above 37°C.
4. Name the target organ for insulin and describe its effect.
5. Name two female sex hormones.

**Key Words**: Hormone • Receptor • Effector

# Coordination and Response in Plants

## Receptors and Effectors in Plants

Plants, as well as animals, have **receptors** and **effectors** so that they can respond to stimuli.

The **receptors detect the stimuli** such as changes in the direction of light or the direction of gravity. The cells of the plant respond by making the plant grow towards or away from the light or gravity.

## Plants Respond to Stimuli

Plant **hormones** are chemicals that control:
- the growth of shoots and roots
- flowering and the ripening of fruits.

Receptors in the plant detect changes in the environment and cause plant hormones called **auxins** to move through the plant in solution. The auxins affect the plant's growth by responding to **gravity (geotropism)** and **light (phototropism)**.

**Shoots** grow:
- towards light (positive phototropism)
- against gravity (negative geotropism).

**Roots** grow:
- away from light (negative phototropism)
- downwards in the direction of gravity (positive geotropism) to absorb water and provide support for the plant.

Growth towards light increases the plant's chance of survival as it can get light for photosynthesis.

**Auxin** is made in the shoot **tip**. Its distribution through the plant is determined by light and can, therefore, be unequal. This is what happens when light shines on a shoot:

1. The hormones in direct sunlight are destroyed.
2. The hormones on the shaded side continue to function, causing the cells to elongate (lengthen).
3. The shoot bends towards the light.

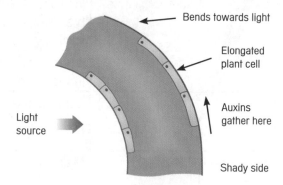

**Experiment to Show that Shoots Grow Towards Light**

1. Cut a hole in the side of a box. Put three seedlings into the box. The seedlings detect light coming from the hole, and will grow towards it.
2. Cut a hole in the side of another box. Put three seedlings with foil-covered tips in the box. These shoots can't detect the light so they grow straight up.

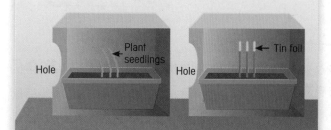

## Quick Test

1. Which parts of a plant respond by:
   a) positive geotropism?
   b) negative phototropism?
   c) positive phototropism?
2. Explain how a shoot tip is made to grow towards light.
3. Explain why phototropism helps plants to grow.

**Key Words**  **Auxin**

# Nervous Coordination in Humans

## Carrying Information Around the Body

Hormones and nerves both carry information around the body, but there are differences between them.

| Nerves | Hormones |
| --- | --- |
| Carry electrical impulses to effectors | Carry information to target organs by chemical messengers |
| Impulses travel very quickly to target organs | Chemical information travels slowly to target organs |
| Response is very rapid and short term | Response is long term |
| Nerves act in a precise area | Effects of hormones can be all over the body |

## Organisation of the Nervous System

A **stimulus** is a change in an organism's environment. Animals respond to **stimuli** in order to keep themselves in suitable conditions for survival.

An animal's response is coordinated by the **central nervous system (CNS)**. This part of the system is sometimes referred to as the **processing centre**.

The CNS (brain and spinal cord) is connected to the body by the **peripheral nervous system (PNS)**.

The peripheral nervous system consists of:
- **sensory** neurones that carry impulses from **receptors** to the CNS
- **motor neurones** that carry impulses from the CNS to **effectors**.

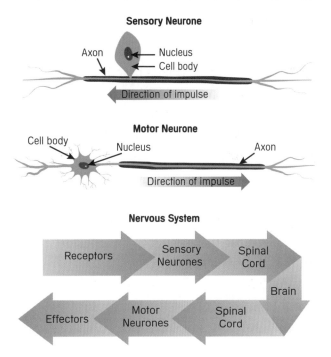

## Types of Receptor

Cells called **receptors** in your sense organs detect **stimuli** (changes in your environment). Different stimuli are detected by different receptors:
- Receptors in your **eyes** are sensitive to **light**.
- Receptors in your **ears** are sensitive to **sound** and **changes in position**, which help your balance.
- Receptors in your **nose** and on your **tongue** are sensitive to chemicals, helping you **smell** and **taste**.
- Receptors in your **skin** are sensitive to **touch**, **pressure**, **pain** and **temperature** changes.

Light receptor cells, like most animal cells, have a nucleus, cytoplasm and cell membrane.

## Quick Test

1. State two differences between nervous and hormonal systems of communication.
2. What type of neurone carries impulses from the central nervous system to an effector such as a muscle?

**Key Words**  Central nervous system • Neurone

# Nervous Coordination in Humans

## Neurones

The central nervous system is made up of the brain and spinal cord. It is linked to sense organs and effectors by nerves (neurones).

Neurones are **specially adapted cells** that carry **electrical signals**, i.e. nerve impulses. There are three types of neurone, each with a slightly different function.

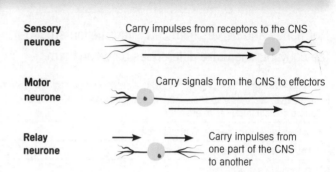

## Reflex Actions

**Reflex actions** are designed to prevent your body from being harmed.

**Reflex actions** are **automatic** and **quick**. They speed up your response time because they by-pass your brain. They involve sensory, relay and motor neurones. Your spinal cord acts as the **coordinator**.

1. Your **receptor** cells detect a stimulus.
2. An impulse travels along a **sensory** neurone.
3. The impulse is passed along a **relay** neurone in the spinal cord in the **CNS**.
4. The impulse travels along a **motor** neurone.
5. The impulse reaches the organ (**effector**) that brings about a response.

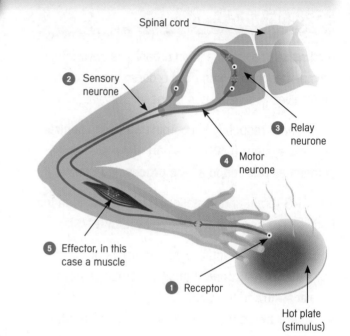

## The Eye

The eye is an organ which acts as a receptor. Receptor cells in the eye detect the stimulus – light. Structures in the eye direct and focus the light onto the receptor cells in the retina. The amount of light entering the eye must be controlled – too much can damage the receptor cells; too little and the receptor cells will not be activated.

The **iris** (the coloured part of the eye) controls the amount of **light** that enters your eye. The rays of light are **refracted** by your **cornea**, and the **lens** focuses light onto the **retina** so the rays **converge** and produce a clear **image** on the retina. The light-sensitive receptor cells cause nerve impulses to pass via the optic nerve to the brain.

The retina contains the light-sensitive receptors. Some are sensitive to colour.

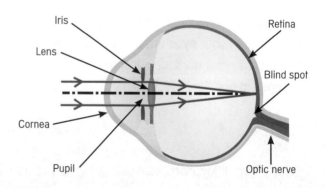

**Key Words**: Reflex action • Retina

# Nervous Coordination in Humans

## Focusing the Light

The eye **lens** is a clear, flexible bag of fluid surrounded by circular **ciliary** muscles that change the shape of the lens (accommodation). **Suspensory** ligaments attach the lens to the ciliary muscles.

When receiving light rays from a near object:
- the ciliary muscles contract
- the suspensory ligaments relax
- the lens is short and fat to refract light a lot.

When receiving light rays from a distant object:
- the ciliary muscles relax
- the suspensory ligaments contract
- the lens is long and thin because the light only needs to be refracted a little.

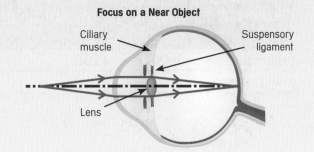

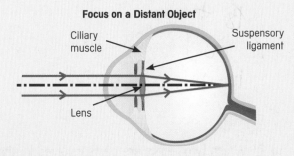

## Responding to Changes in Light Intensity

The **iris** is a ring-shaped disc of muscle that is the coloured part of the eye.

The hole in the centre of the ring is the pupil. The iris is made of **circular and radial muscles** that can contract or relax to change the size of the pupil.

In **dim light** conditions, the iris muscles widen the pupil. This allows more light to reach the receptor cells in the retina. **The pupil is dilated**.

When **light is bright** the iris muscles make the pupil smaller to protect the receptor cells from damage. The **pupil is constricted**.

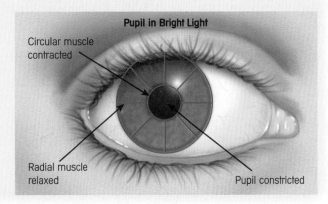

## Quick Test

1. Explain why reflex actions are important.
2. In which part of the eye are light receptor cells found?
3. Which parts of the eye help to focus light?
4. P2 How is the lens pulled thin when you look at a distant object?
5. P2 Which muscle in the eye contracts when the light is dim?

# Exam Practice Questions

**1** The diagram shows a kidney tubule from a normal, healthy person.

**A Kidney Tubule**

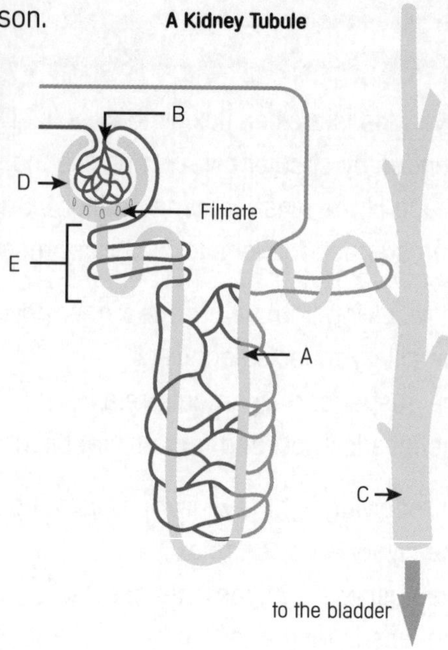

| Substance in Blood | Amount (g/day) | | |
|---|---|---|---|
| | Blood in Renal Artery | Filtrate | Urine |
| Glucose | 1000 | 200 | 0 |
| Blood proteins | 80 000 | | 0 |

**a)** Name the following parts of the diagram:

A: *Glomular*

B: *B* ~~⟍~~

C: *Duct*

D: *Bowmans Capsule*

E: ........................................................ [5]

**b) i)** Which two letters are parts involved in ultrafiltration? *B + D* [2]

**ii)** Which process occurs in part C? [1]

*Water is reabsorbed to blood*

**c)** Complete the table to show the amount of protein found in the filtrate. [1]

**d)** Explain why blood protein is not normally found in urine. [1]

*Learn.*

**e)** Explain why glucose is found in the filtrate but is not found in the urine. [3]

# Exam Practice Questions

**2** The presence of protein and glucose in urine is a sign of illness. Protein is sometimes found in the urine of people who have high blood pressure. Glucose in the urine can indicate diabetes.

   **a)** Describe how you could use Benedict's Reagent to test a sample of urine for the presence of glucose. **[2]**

   **b)** Suggest why protein appears in the urine of people with high blood pressure. **[2]**

**3** The apparatus shown can be used to measure the rate of transpiration in a plant.

   **a)** What is transpiration? **[2]**

   *loss of water vapour from a leaf*

# Exam Practice Questions

**b)** A pupil recorded the mass of the plant every 30 minutes for 4 hours. The pupil noted that after an hour the Sun came out and made the lab that the plant was in very warm and sunny. The results are shown in the table.

| Time from Start (mins) | Mass of Plant (g) |
| --- | --- |
| 0 | 240 |
| 30 | 235 |
| 60 | 230 |
| 90 | 215 |
| 120 | 195 |
| 150 | 185 |
| 180 | 160 |
| 210 | 140 |
| 240 | 120 |

**i)** Calculate the total mass lost in 240 minutes. [1]

*120*

**ii)** Calculate the mass lost per hour over the whole of the experiment. [2]

*35/hr*

**c)** On the grid, plot the data to show the mass of the plant during the four hours. Label the axes. Join your points using straight lines. [5]

**d)** Explain why the plant lost mass. [1]

*Water is evaporating from leaves, then xylem is replacing it, but slowly running out of water*

# Exam Practice Questions

e) Describe the change in mass shown on the graph. [2]

*No indeed* (handwritten)

f) Explain why the rate of transpiration changes after the first hour of the experiment. [4]

**4** Water vapour is lost through the stomata of plant leaves.

a) Name two other gases that pass into or out of the plant leaves through the stomata. [1]

*$CO_2$, $O$* (handwritten)

b) Which processes produce these gases? [2]

*Photosynth, Respiration* (handwritten)

c) Describe how the structure of a leaf is adapted for efficient gas exchange. [4]

# Reproduction

## Types of Reproduction

The two types of reproduction are:
- **sexual reproduction**
- **asexual reproduction**.

During **sexual reproduction** male and female **gametes** fuse together. This is called **fertilisation**.

The genes carried by the egg and the sperm are mixed together to produce a new individual. This process produces lots of **variation**, even amongst offspring from the same parents.

When the egg fuses with the sperm a **zygote** is formed. The zygote then divides into many different cells to make an **embryo**.

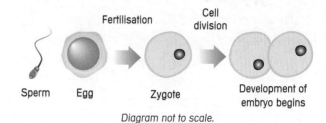

*Diagram not to scale.*

**Asexual reproduction** doesn't produce any genetic variation at all. Only one parent is needed, so there is no mixing of genes. All offspring are genetically **identical** to the parent, i.e. **clones**. Variation can only be due to environmental factors.

## The Human Reproductive System

The diagrams show the key parts of the female and male reproductive systems.

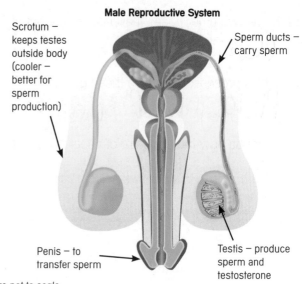

*These diagrams are not to scale.*

## The Menstrual Cycle

During the **menstrual cycle**, the uterus lining has different thicknesses. There are four stages:
1. The uterus lining breaks down (called a period).
2. The uterus wall is repaired and gradually thickens.
3. An egg is released from an ovary (**ovulation**).
4. The uterus lining stays thick in preparation for a **fertilised** egg. If no fertilised egg is detected, the cycle starts again.

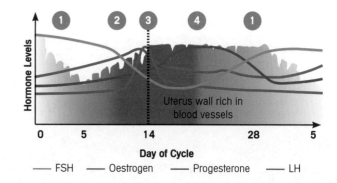

**Key Words**   Sexual reproduction • Asexual reproduction • Fertilisation • Zygote • Embryo • Menstrual cycle

# Human Reproduction

## The Menstrual Cycle (Cont.)

FSH (follicle-stimulating hormone) is a hormone that stimulates the egg to ripen in the ovary. The ovary releases **oestrogen**, a **hormone that stimulates the uterus lining to thicken**, and stimulates the release of LH (luteinising hormone).

LH is the hormone that causes **ovulation** about halfway through the menstrual cycle. After ovulation, **progesterone** is produced by the ovary to **preserve the uterus lining**. FSH and LH are released by the pituitary gland in the brain.

As oestrogen and progesterone levels fall towards the end of the cycle, **menstruation** occurs (i.e. the uterus lining breaks down).

## P2 Fertilisation and the Placenta

During sexual intercourse sperm are deposited in the vagina and swim towards the ovaries. If sperm meet an ovum in the oviduct, fertilisation takes place.

The fertilised ovum begins to develop into an embryo. It implants in the wall of the uterus and forms the placenta, which separates the blood of the mother from the blood of the foetus.

The placenta lets the blood of the mother and the embryo flow close together to allow the exchange of nutrients, oxygen and waste products:

- **Oxygen and nutrients and antibodies** diffuse from the mother's blood to the foetus.
- **Carbon dioxide and urea** diffuse from the foetus back to the mother.
- The placenta helps **to prevent pathogens and some harmful chemicals** getting into the foetus.

The developing foetus is joined to the placenta by the umbilical cord.

After implantation the amnion, a membrane, forms and surrounds the embryo. The amnion secretes amniotic fluid **which protects the developing foetus from knocks and bumps**.

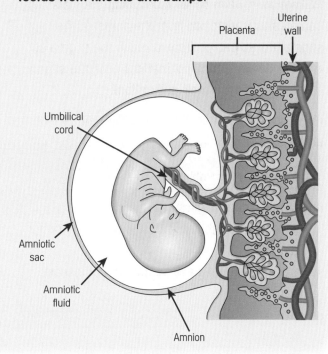

## Quick Test

1. What is the difference between sexual and asexual reproduction?
2. What is a zygote?
3. Where are male and female gametes produced in humans?
4. P2 Explain what happens in the placenta.
5. P2 What is the function of amniotic fluid?

**Key Words** — Oestrogen • Progesterone • Placenta • Foetus • Umbilical cord • Amniotic fluid

# Plant Reproduction

## Sexual Reproduction in Flowering Plants

**Male gametes are inside pollen grains** which are produced by the stamens of the flower. **Female gametes are inside ovules** that are made in the ovary of the plant.

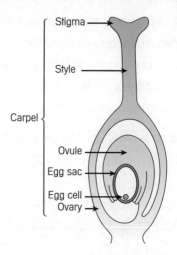

**Carpels** are the female parts of the flower. Carpels are made up of **stigma, the style and the ovary**.

**Stamens** are the male parts of the flower. Stamens produce pollen, which contains the male gamete, in the anther at the top. When the anther ripens the pollen grains are released

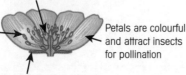

Petals are colourful and attract insects for pollination

Sepals protect the flower when it is a bud

## Pollination

Pollination is when pollen (containing the male gametes) is transferred from the anther to the stigma. Pollen can be transferred to the stigma of **another flower (cross-pollination)**, or to the stigma of the **same flower (self-pollination)**.

Pollen is carried to another flower by wind or by animals, especially insects.

### Adaptations for Pollination by Insects

Insect-pollinated flowers:
- have **large brightly coloured petals** to attract insects
- are **scented** and **have nectaries** that insects feed on
- have large sticky or spiky pollen grains that stick to insects
- have sticky stigmas so that pollen on the insect sticks to the stigma as the insect brushes past it.

### Adaptations for Wind Pollination

Wind-pollinated flowers:
- have small, dull petals (they don't need to attract insects)
- have long filaments with anthers that hang outside the flower so the pollen is easily blown off by the wind
- produce large quantities **of tiny, light, smooth pollen grains** that are easily carried by the wind
- have **large feathery stigmas** hanging outside the flower, which can catch pollen blown in by the wind.

**Wind Pollinated Flower**

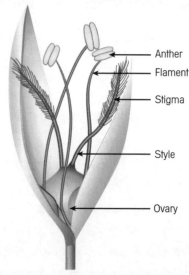

Key Words: **Pollen • Ovule**

# Plant Reproduction

## Fertilisation in Flowering Plants

1. The pollen grain lands on the stigma and then starts to grow a **pollen tube** down the **style to the ovary**.
2. The end of the pollen tube reaches the **micropyle**.
3. The male gamete passes down the tube into the ovule.
4. The male gamete nucleus fuses with the ovule nucleus, making a zygote.

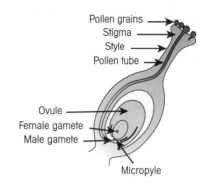

## Seeds and Fruits

The zygote divides into many cells that form the embryo plant. The embryo plant is made of:
- the embryo shoot (**plumule**)
- the embryo root (**radical**)
- the embryo leaves (**cotyledons**).

The **cotyledons store food**. The outside of the ovule hardens to form the seed coat – **testa**.

Ovary tissue around the seed sometimes develops into the **flesh of a fruit**.

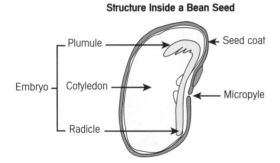

## Germination of Seeds

To germinate, seeds need: **water, oxygen** and **warmth**.

If they don't have the right conditions they may stay dormant. Germination occurs in the following stages:

1. Water enters the seed and **activates enzymes** that can break down stored food into soluble nutrients.
2. Warmth **increases the rate of the enzyme action**.
3. The **nutrients diffuse** to the embryo root and shoot.
4. Cells in the embryo use the nutrients and oxygen to **respire and grow**.
5. Once leaves are formed above ground the young plant can begin to **photosynthesise**. To do this the young plant needs light.

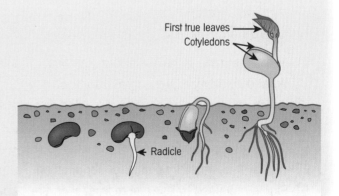

**Key Words**: Pollen tube • Micropyle • Plumule • Radical • Cotyledon • Testa

# Plant Reproduction

## Asexual Reproduction in Plants – Natural Methods

**Asexual reproduction** produces identical copies.

Plants can reproduce asexually, i.e. in the absence of sex cells and fertilisation.

Spider plants, strawberry plants and potato plants all reproduce in this way.

Spider Plant Runners

Runner – a rooting side branch | New individual established | New individual (genetically identical) now independent

## Asexual Reproduction in Plants – Artificial Methods

### Taking Cuttings

Plants grown from cuttings or tissue culture are **clones**. If a plant has desirable characteristics, it can be reproduced by taking stem, leaf or root cuttings. (See also p. 94.)

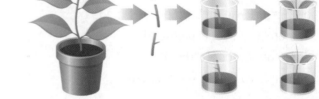

Select a plant | Take cuttings | Place in damp atmosphere | New genetically identical plants develop

### Cloning by Tissue Culture

Cloned plants can be produced by the following method:

1. Select a parent plant with desired characteristics.
2. Scrape off a lot of small pieces of tissue into beakers containing nutrients and hormones. Make sure that this process is done **aseptically** (without the presence of bacteria or fungi) to avoid the new plants rotting.
3. Lots of genetically identical plantlets will then grow (these can also be cloned).

Many older **plants** are still able to **differentiate** or **specialise**, whereas animal cells lose this ability. So cloning plants is easier than cloning animals.

## Quick Test

1. Name the parts of the flower that produce: **a)** male gametes **b)** the female gamete.
2. What do we mean by the term pollination?
3. List three features of: **a)** wind-pollinated flowers **b)** insect-pollinated flowers.
4. Describe how the male gamete reaches the ovule from the stigma of a plant.
5. Name: **a)** the embryo shoot **b)** the embryo root **c)** the embryo leaves **d)** the seed coat.
6. P2) List three factors needed for germination.
7. Plants produced by asexual reproduction are clones of their parent. What does this mean?

**Key Words**: Clone

# Inheritance

## Chromosomes, Genes and DNA

Inside the nucleus of every cell are **chromosomes**. Chromosomes are long coiled molecules of DNA, divided up into regions called **genes**, also made of DNA.

The genes carry information in the form of a genetic code. The code is the sequence of bases found along the DNA molecule. Each gene has a different sequence of bases.

Each gene is the genetic instruction that controls a particular inherited characteristic.

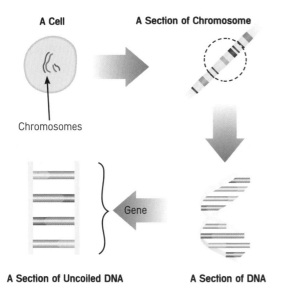

Genes control a particular characteristic by providing a code for a combination of amino acids that make up a specific protein.

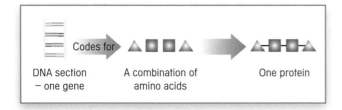

The nucleus of each cell contains a complete set of genetic instructions, carried by genes on the chromosomes. The instructions control how the organism is constructed and how the individual cells work. Most body cells have the same number of chromosomes, in matching pairs.

### DNA

**DNA** controls the production of proteins. Proteins are needed for growth and repair. A gene is a length of DNA. Genes are like recipes for proteins – each gene codes for a particular protein.

A DNA molecule is made of two strands coiled around each other in a **double helix** (spiral). The genetic instructions are in the form of a chemical code made up of four **bases**. These **bases** bond together in specific pairs, forming cross-links. Each gene contains a different sequence of bases.

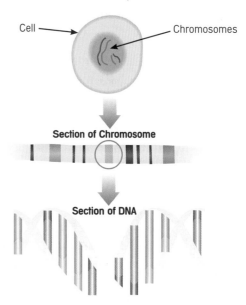

The four bases in DNA are **A**, **C**, **G** and **T**.

On opposite strands of the DNA molecule:
- A always bonds with T
- C always bonds with G.

This is complementary 'base pairing'.

**Key Words**: Chromosome • Gene • Double helix

# Inheritance

## Human Body Cells

Body cells contain **two sets** of **23 chromosomes** arranged in pairs (46 in total). These cells are **diploid**.

Chromosomes contain genetic information.

**Gametes** are sex cells, i.e. female eggs and male sperm. Gametes only have **one set** of **23** chromosomes. These cells are **haploid**.

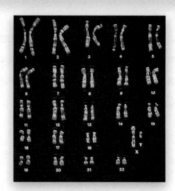

## Fertilisation

During **fertilisation**, the female and male gametes fuse to produce a **zygote** – a **single body cell** with **23 new pairs** of chromosomes.

In each pair:
- one chromosome comes from the mother
- one chromosome comes from the father.

The cell then divides repeatedly by **mitosis** to form a new individual.

Variation is caused due to the different combination of genes from the mother and father.

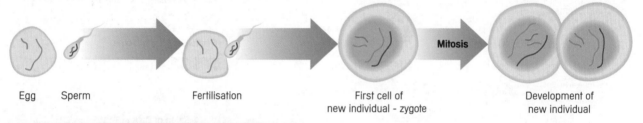

Egg   Sperm            Fertilisation            First cell of new individual - zygote            Development of new individual

## Alleles

Some characteristics are controlled by a single **gene**.

**Genes** may have different forms, or variations, called **alleles**.

For example:
- the gene that controls tongue-rolling ability has two alleles – either you can or you can't
- the gene that controls eye colour has two alleles – blue or brown.

**Sexual reproduction** gives rise to **variation** because when gametes join during fertilisation:
- one allele for each gene comes from the mother
- one allele for each gene comes from the father.

In a pair of **chromosomes**, the alleles for a gene can be the **same** or **different**. If they are different:

- one allele will be **dominant**
- one allele will be **recessive**.

A **dominant** allele **will always control** the characteristic; it will express itself even if present on only one chromosome in a pair.

A **recessive** allele will **only** control the characteristic if it is present on **both chromosomes in a pair** (i.e. no dominant allele is present).

### Quick Test

1. What is: **a)** a chromosome? **b)** a gene?
2. How do bases pair in DNA?
3. Explain what we mean by a dominant allele.

**Key Words**  Gametes • Mitosis • Alleles • Dominant allele • Recessive allele

# Inheritance

## Combinations of Alleles

The diagram shows three pairs of alleles on the middle of a pair of chromosomes. The alleles code for:
- tongue-rolling ability
- eye colour
- type of earlobe (i.e. attached or unattached).

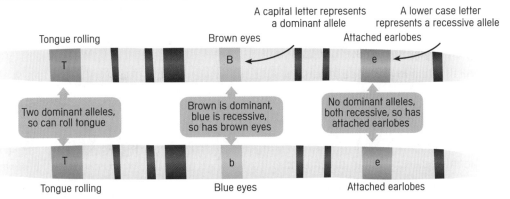

| Characteristic | Homozygous Dominant | Heterozygous | Homozygous Recessive |
|---|---|---|---|
| Tongue rolling | TT (can roll) | Tt (can roll) | tt (can't roll) |
| Eye colour | BB (brown) | Bb (brown) | bb (blue) |
| Earlobes | EE (free earlobes) | Ee (free earlobes) | ee (attached earlobes) |

## Co-dominance

Sometimes alleles are co-dominant. Both the alleles are **equally dominant, so characteristics of both alleles are present** in the organism.

**Example 1:**
R is the allele for red flowers in snapdragons, r is the allele for white flowers:
- RR flowers are red.
- rr flowers are white.
- Rr flowers are pink.

**Example 2:**
Human blood type is controlled by three alleles A, B and O:
- A and B alleles are both **dominant to O**.
- **A and B alleles are also co-dominant** with each other.

The blood group that a person has depends on a different combination of these alleles:
- AO or AA gives a person blood group A because A is dominant to O.
- BO or BB gives a person blood group B because B is also dominant to O.
- OO gives a person blood group O.
- But AB gives the person a different blood group – group AB (because groups A and B are co-dominant).

**Key Words** — Co-dominant

# Inheritance

## Monohybrid Inheritance

When a characteristic is determined by **just one pair** of **alleles**, it is referred to as **monohybrid inheritance**. A simple **genetic diagram** is a biological model and can be used to predict the outcome of crosses in monohybrid inheritance.

### Genetic Diagrams

In genetic diagrams, you should use **capital** letters for **dominant** alleles and **lower case** letters for **recessive** alleles. So, for eye colour,
- B is used for brown eye alleles (dominant)
- b is used for blue eye alleles (recessive).

When constructing genetic diagrams, remember:
- to clearly identify the alleles of the parents
- to place each of these alleles in a separate gamete
- to join each gamete with the **two gametes** from the other parent.

You should know the following genetic terms:
- **Genotype** – the combination of alleles that an individual has for a particular gene.
- **Homozygous** – an individual who carries two copies of the **same** allele for a particular gene, e.g. **BB** or **bb**.
- **Heterozygous** – an individual who carries two **different** alleles for a particular gene, e.g. **Bb**.
- **Phenotype** – the expression of the genotype (the characteristic shown), e.g. the **homozygous recessive genotype** of bb would have a **phenotype** of blue eyes.

From the crosses on the diagrams, the following can be seen:

**1** If one parent has **two dominant** alleles, then **all offspring** inherit the characteristic.

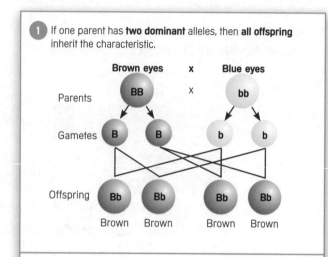

**2** If two parents have **one recessive allele each**, then the characteristic **may** appear in offspring (25% chance).

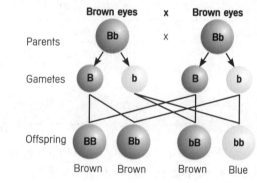

**3** If one parent has **one recessive** allele and one parent has **two recessive** alleles, then there is a 50% chance that the characteristic will appear in offspring.

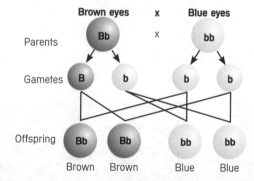

Remember, these are **only probabilities**. In practice, what matters is which egg is fertilised by which sperm. This process is completely **random**.

### Quick Test

1. What are different forms of genes called?
2. What is the difference between a phenotype and a genotype?
3. How do we describe the genotype of a person whose cells have both recessive and dominant alleles of the same gene?
4. P2 In human blood groups which alleles are: **a)** recessive **b)** co-dominant?

**Key Words**  Genotype • Homozygous • Heterozygous • Phenotype

# Inheritance

## Using Family Pedigrees to work out Genotypes

Recessive characteristics only show in individuals who **have two recessive alleles**. For example, blonde hair (b) is recessive to brown (B) and is controlled by a single gene. Blonde hair only appears where the genotype is homozygous recessive (bb).

| Genotype | Phenotype |
|---|---|
| BB homozygous dominant | Brown hair |
| Bb heterozygous | Brown hair |
| bb homozygous recessive | Blonde hair |

Look at this family pedigree for hair colour:

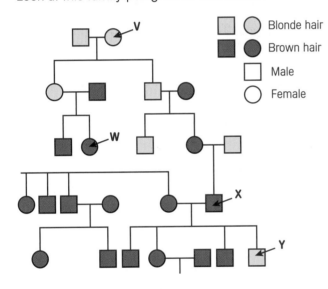

The phenotypes shown are:
- V = blonde-haired female
- W = brown-haired female
- X = brown-haired male
- Y = blonde-haired male.

The genotypes must therefore be as follows:
- **V = bb** because the **blonde phenotype is recessive**, as any dominant brown allele (B) would have shown up as brown hair in the individual.
- **W has brown hair and so must have at least one dominant allele (B)**. W must be Bb (heterozygous) because her mother is blonde (bb). Therefore W has one recessive allele (b) from her mother and a dominant allele (B) from her brown-haired father.
- **X = Bb** because he is married to a brown-haired female and **between them they have a blonde-haired child** (Y = bb). The child must have inherited a recessive (b) allele from each parent, i.e. **both parents must be heterozygous**.

Family trees like this have been used to explain how genetic disorders like cystic fibrosis, Huntingdon's disease and haemophilia are inherited.

## Inheritance of Sex

**Gender** (in mammals) is determined by the **sex chromosomes**: XX = female; XY = male.

**Egg cells** all carry X chromosomes. Half of **sperm cells** carry X chromosomes and half carry Y chromosomes.

The sex of an individual depends on whether the egg is **fertilised** by an X- or Y-carrying sperm:
- An egg fertilised by an X sperm will become a girl (X from egg and X from sperm = XX).
- An egg fertilised by a Y sperm will become a boy (X from egg and Y from sperm = XY).

The **chances** of an egg being fertilised by an X-carrying sperm or a Y-carrying sperm are equal, so there are approximately equal numbers of male and female offspring.

This genetic diagram shows a 50:50 chance of having a boy or girl:

|  |  | Male | |
|---|---|---|---|
|  |  | X | Y |
| Female | X | XX | XY |
|  | X | XX | XY |

# Inheritance

## Mitosis

Mitosis is the division of body cells to produce new cells.

Mitosis occurs:
- for growth
- for repair
- for cloning
- in asexual reproduction (cells produced by asexual reproduction contain the same alleles as the parents).

During mitosis, the following takes place:
1. A copy of each chromosome is made.
2. The cell then divides **once** to produce two body cells.
3. The **new cells** contain exactly the **same genetic information** as the **parent** cell, i.e. the same number of chromosomes and the same genes.

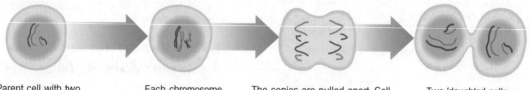

Parent cell with two pairs of chromosomes → Each chromosome replicates itself → The copies are pulled apart. Cell now divides for the only time → Two 'daughter' cells are formed

## Meiosis

**Meiosis** is a type of cell division which occurs in the testes and ovaries. The cells in these organs divide to produce **gametes** for sexual reproduction.

When a **diploid** cell divides by meiosis, four new cells are produced. In each cell formed the chromosome number is halved, so the cells are **haploid**, and each cell is genetically different. Meiosis introduces genetic **variation**. In human cells the diploid number is 46 and the haploid number is 23.

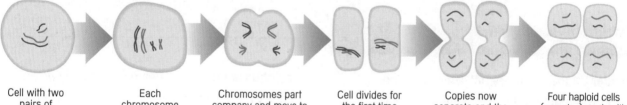

Meiosis – the cell divides twice to produce four cells with genetically different sets of chromosomes

Cell with two pairs of chromosomes (diploid cell) → Each chromosome replicates itself. → Chromosomes part company and move to opposite poles. → Cell divides for the first time. → Copies now separate and the second cell division takes place. → Four haploid cells (gametes), each with half the number of chromosomes of the parent cell.

### Quick Test

1. What is the genotype of a male?
2. What percentage of all offspring are likely to be male?
3. How is mitosis used in organisms?
4. Give three differences between mitosis and meiosis.
5. Explain how meiosis gives rise to genetic variation.

**Key Words**  Meiosis

# Inheritance

## Variation

Differences between individuals of the same species are called **variation**. Variation can be due to:
- **genetic factors**, e.g. dimples, eye colour
- **environmental factors**, e.g. scars, hairstyle
- **a combination** of both factors, e.g. weight may be due to diet as well as the genes you inherit.

**Meiosis** and **sexual reproduction** produce **variation** between offspring and parents:
- When the gametes fuse, genetic information from two individuals is combined.
- For each gene, just one of each parent's alleles is passed on.
- Each offspring can have a different combination of alleles from either parent.
- The offspring have different characteristics from each other.

Genetic Causes

Environmental Causes

## Mutations

**Mutations** are rare, random changes in the genetic material that can be inherited. **Most** mutations are **harmful** or neutral, although **occasionally** a **beneficial mutation** occurs.

**P2** Gene **mutations** are changes to **genes**. These changes can be spontaneous but the rate can be increased by ionising radiation, e.g. gamma rays, X-rays and ultraviolet rays, and some chemical mutagens, such as substances in tobacco.

Mutations may lead to production of different proteins. Mutations change the base sequence of **DNA**. This alters the shape and function of the protein or prevents the production of the protein that the gene normally codes for.

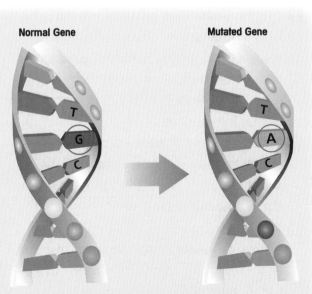

The G base is substituted for an A base

**Key Words**  Variation • Mutation

# Inheritance

## The Theory of Evolution

Animals and plants that are better adapted to their environment are more likely to survive. This theory is called **natural selection** and was first put forward by **Charles Darwin**.

These adaptations are controlled by **genes** and can be **passed on** to future generations.

If the environment changes, the best adapted individuals of the species survive and reproduce. They pass on the genes for the favourable adaptations to their offspring.

Individuals in a species that aren't well adapted to their environment may become **extinct**.

A Dodo

## Evolution by Natural Selection

After making extensive observations, **Charles Darwin** proposed his theory of **evolution by natural selection**.

Evolution is the **specialisation** of a population over many generations to become better **adapted** to its environment. There are four key points to remember:

1. Individuals within a population show **natural variation** (i.e. differences due to their genes that may arise through mutations).
2. There is **competition** between individuals for limited resources (e.g. food, mates) and also predation and disease, which keep population sizes constant in spite of the production of many offspring, i.e. there is a 'struggle for survival' and weaker individuals die.
3. Individuals that are **better adapted** to the environment are more likely to **survive**, breed successfully and produce offspring. This is termed '**survival of the fittest**'.
4. These survivors will therefore pass on their 'successful' **genes** to their **offspring**, resulting in an improved organism being evolved through natural selection.

Groups of the same species who are separated from each other by physical boundaries like mountains or seas will not be able to breed and share their genes. This is because over long periods of time the separate groups may specialise so much that they can't successfully breed any longer, and so two new species are formed.

As new discoveries have been made, and a better understanding of genetics and inheritance is known, the theory of natural selection has been developed and updated.

Many theories have been put forward in the past to explain how evolution occurs. The theory of evolution by natural selection was initially met with hostility. Darwin's ideas went against those of the Church and the Bible. Most scientists now accept the theory put forward by Darwin but there is still debate amongst some scientists about the precise mechanism.

# Inheritance

## Examples of Natural Selection Today

### Peppered Moths

**Peppered moths** can be pale or dark. Pale peppered moths are easily camouflaged amongst the lichens on silver birch tree bark.

But, in areas of high pollution, the bark on silver birch trees was discoloured by soot. In these areas, the rarer, darker speckled varieties of the peppered moth were more common than the pale varieties.

This is because the pale peppered moths show up against the sooty bark, whereas the darker peppered moths are camouflaged. So they were able to survive and breed in greater numbers.

### Bacteria and Penicillin

Some populations of **bacteria** have become **resistant** to penicillin by natural selection:

1. Bacteria mutated. Some were resistant to the antibiotic penicillin; others were not.
2. The non-resistant bacteria were more likely to be killed by the penicillin.
3. The penicillin-resistant bacteria survived.
4. The surviving bacteria reproduce, leading to more bacteria in the population that are resistant to penicillin.

As resistance to antibiotics spreads in the populations of bacteria it becomes more difficult to treat diseases caused by the bacteria, because the antibiotics do not work. This is why doctors are reluctant to prescribe antibiotics unless they're absolutely necessary.

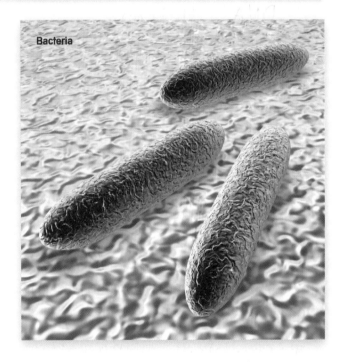

## Quick Test

1. What is a mutation?
2. Who put forward the theory of evolution through natural selection?
3. What is natural selection?
4. Why shouldn't we use antibiotics unless absolutely necessary?
5. Why was Darwin's theory unpopular with the Church of England at the time?

# Exam Practice Questions

1.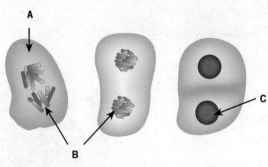

a) The diagrams show a cell dividing to form two cells. What kind of cell division is taking place? [1]

Mitosis

b) Name structures: [3]

A: Cell

B: Chromosomes

C: Nucleus

c) What chemical is structure B made from? [1]

DNA + Histones

d) The two cells formed are clones. Explain what this means. [2]

Diploid is exactly the same as parent cell, except for possible mutations.

e) The two cells that are produced are both diploid. What does this mean? [1]

Daughter clone?

f) How many chromosomes are there in a human sperm cell? [1]

23

g) A different kind of cell division occurs to make gametes. Describe three ways in which cell division to make gametes is different from the cell division shown in the diagram. [3]

Not clone. Half chromosomes.

# Exam Practice Questions

**2** Sexual reproduction in plants involves pollination. The transfer of pollen between two flowers of the same species is shown in the diagram.

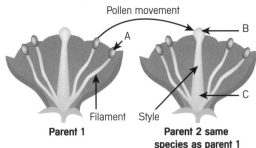

a) Identify: [3]

   A: ................................................................

   B: ................................................................

   C: Ovary

b) Suggest how pollen is transferred from A to B. [1]

   Bees, Wind.

c) Describe one feature shown in the diagram that is an adaptation to this kind of pollination. [1]

   ................................................................................................................

   ................................................................................................................

   ................................................................................................................

d) In this species there are red and also white varieties of flower. Colour is controlled by a single gene. Red is a dominant allele (R). White is recessive (r).

   Parents 1 and 2 are both heterozygous for petal colour.

   i) Complete the spaces and diagram to show the genotypes of the parents and their possible offspring. [5]

   Parent 1 phenotype ............................... × Parent 2 phenotype ...............................

   Parent 1 genotype ............................... × Parent 2 genotype ...............................

   Gametes ............... ............... × ...............

   | Genotypes |   |   |          |
   |-----------|---|---|----------|
   |           |   |   | Offspring |
   |           |   |   |          |
   
   Offspring

   ii) What proportion of the offspring is likely to have white flowers? [1]

   ................................................................................................................

# The Organism in the Environment

## Ecological Terms

An **ecosystem** is a physical environment with a particular set of conditions, plus all the organisms that live in it. An ecosystem can be natural or artificial.

A **habitat** is the part of the physical environment where an animal or plant lives. An organism will have adapted to its habitat, so it may be restricted to living there. It may only eat the food in that location.

A **population** is the total number of individuals of the same species that live in a certain area.

A **community** is the total number of individuals of the different populations of organisms that live together in an area at the same time.

## Population Size and Distribution

There are a number of **physical factors** that affect the **number of each species** found in an environment. These include:
- temperature
- the availability of water
- the availability of light
- the availability of nutrients
- the availability of oxygen and carbon dioxide.

These factors also affect where the organisms are found, in other words **how they are distributed**.

## Sampling Methods

It is normally very difficult to count **all the species in a habitat, so samples** are taken. The size of the sample affects how valid and reproducible the data is.

**Quantitative** data on the distribution of organisms can be obtained in two ways: **random sampling with quadrats** and **sampling along a transect**.

### Random Sampling with Quadrats

A student wants to know how many of each different species of plants there are in her school field.

1. The student sets out a sample area of the school field (e.g. 100m²).
2. A 1m² **quadrat** is placed randomly in the sample area.
3. The number of each different plant species found in the quadrat is recorded.
4. This is **repeated** several times.
5. A **mean** is calculated from the data by adding up the number of each species found in each quadrat and dividing by the number of quadrats used.
6. The total number of organisms in the school field can then be **estimated** by multiplying the mean number by the total area of the school field.

A quadrat can come in different sizes, but 1m² or 0.5m² are most commonly used.

**Key Words**: Ecosystem • Habitat • Population • Community • Quantitative • Quadrat

# The Organism in the Environment

## Sampling Methods (Cont.)

### Sampling along a Transect

A student wants to find out if the shade of a big hedge in his school field affects the number of species of plants that are able to grow near it. The student is investigating how the shade of the hedge affects the distribution of the plants of different species.

1. The student stretches some string from the hedge out into the field, producing a **transect**.
2. A quadrat is placed at 5m intervals along the transect moving away from the hedge.
3. The student records the number of each plant species found in each quadrat along the transect.
4. To increase the **reliability** of the data, the student repeats his investigation by using more transects.

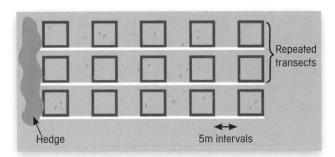

### Using Transects

Counting the animals and plants in quadrats along a transect line gives you lots of numbers. It is easier to see the pattern of distribution of the different species if you compare them using a kite diagram to display the data.

You can create a kite diagram by doing the following:

1. Use graph paper and begin by drawing a sketch of the habitat profile across the bottom, to scale.
2. Draw a horizontal line above this and locate the quadrats – mark a vertical bar at each quadrat location (use 5 squares above and 5 below for Abundant, 4 for Common, 3 for Some, 2 for Few, 1 for only one).
3. Join the tops and bottoms of these bars. 'Not present' will be a point on the horizontal line, so the diagram that results will have a shape something like a kite. That is the profile for one species done.
4. Do the same for the next species, and so on.

Remember, the presence or absence plus abundance of an organism is affected by other organisms in the area, e.g. predators, as well as other physical factors like the tides or water temperature.

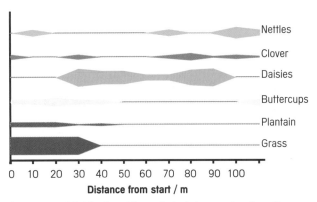

The pattern of distribution of these 6 plants is very clear from the kite graph.

### Quick Test

1. Explain the difference between the terms community and population.
2. Name three physical factors that affect the number of organisms in a particular habitat.
3. Describe how to use a quadrat to sample the number of organisms in a habitat.
4. Suggest how a hedge could affect the population of daisies growing near it.
5. What can kite diagrams show you about populations of organisms?

**Key Words** — Transect

# Feeding Relationships

## Food Chains

**Food chains** show the **transfer of energy** from organism to organism. Energy from the Sun flows through a food chain when green plants absorb sunlight to make glucose, and through feeding.

Green plants are **producers** because they produce **biomass** during **photosynthesis** (algae and plankton are other examples of producers).

**Consumers** are organisms which eat other organisms. All other organisms in food chains rely on plants.

In ecology, the **trophic level** is the position or stage that an organism occupies in a food chain.

**Biomass and energy** are lost at every trophic level of a food chain. Materials and energy are lost in an organism's faeces during **egestion**. Energy is lost through **movement** and **respiration**, especially in birds and mammals, and through heat loss and waste (excretion), and so it doesn't go into making new biomass.

Excretory products and uneaten parts can be used as the starting point for other food chains, e.g. dung beetles eating elephant faeces.

Organisms which eat both plants and animals can be both primary and secondary consumers, e.g. humans.

**A Food Chain**

Green plant, producer

Rabbit, primary consumer

Stoat, secondary consumer

Fox, tertiary consumer

## Efficiency of Energy Transfer

The length of a food chain depends on the **efficiency of energy transfer**. In the food chain shown above:
- a fraction of the Sun's energy is captured by the producers
- the rabbits respire and produce waste products – they pass on a tenth of the energy they get from the grass (10%); 90% is lost
- the stoats respire and produce waste products – they pass on a tenth of the energy they get from the rabbits (10%); 90% is lost
- the fox gets the last tiny bit of energy left after all the others have had a share.

Food chains rarely have fourth or fifth degree consumers, as there isn't enough energy to pass on.

## Food Webs

Food chains link up to make **food webs**. If organisms are removed from, or added to, a food web it has a huge impact on all the other organisms. For example, if the seaweed was removed from this food web, the fish and the winkle numbers may go down due to less food. This may then cause seal and lobster numbers to go down.

Organisms that consume dead producers and consumers, e.g. bacteria and fungi, fit into the food web as **decomposers**.

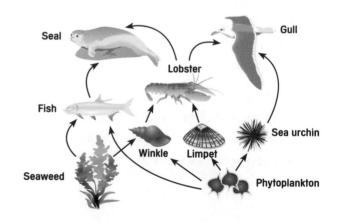

**Key Words**: Producer • Biomass • Consumer • Trophic level • Decomposers

# Feeding Relationships

## Pyramids of Numbers

The number of organisms at each stage in a food chain can be shown as a **pyramid of numbers**.

The number of organisms decreases as you go up the pyramid, i.e. a lot of grass feeds a few rabbits, which feed even fewer stoats, which feed just one fox.

For simplicity, pyramids of numbers usually look like this:

Pyramids of numbers don't take into account the **mass** of the organisms, so it's possible to end up with some odd-looking pyramids. For example, if lots of caterpillars feed on one lettuce, the base of the pyramid is smaller than the next stage. This happens because the lettuce is a large organism compared to the caterpillar. This situation also happens when trees are at the bottom of a food chain.

## Pyramids of Biomass

**Pyramids of biomass** show the dry mass of living material at each stage in the chain. They're usually pyramid shaped because they take the **mass** of the organisms into account.

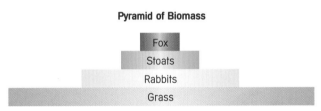

The efficiency of energy transfer **explains the shape of biomass pyramids**. Biomass is lost through the stages. A lot of biomass remains in the ground as the root system. The rabbits and stoats lose biomass in faeces and urine. The fox gets the remaining biomass.

There are problems with creating pyramids of numbers and biomass. Some organisms may belong to more than one trophic level, and measuring biomass is tricky as it involves drying out and weighing the mass of an organism, which isn't easy with large organisms like trees.

## Pyramids of Energy

Pyramids of energy represent the energy in the organisms at each trophic level, found in an area over a period of time. They are **always pyramid shaped**.

The units could be, for example, **kJ/ha/yr**.

### Quick Test

1. List two ways in which energy is lost from a food chain.
2. Approximately what percentage of energy is lost between trophic levels in a food chain?
3. Why is it more difficult to record biomass than number of organisms?
4. Explain why pyramids of numbers can be inverted.
5. What is measured to make a pyramid of energy?

**Key Words**  Pyramid of numbers • Pyramid of biomass

# Cycles Within Ecosystems

## Water Cycle

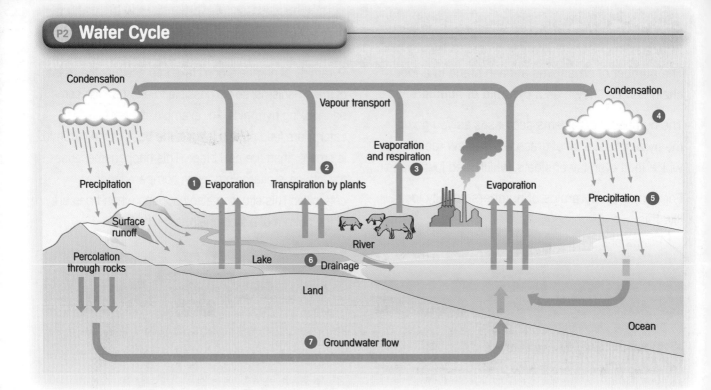

## The Carbon Cycle

In a stable community, materials are constantly being recycled. The constant recycling of carbon is called the **carbon cycle**:

1. Green plants take in carbon dioxide for photosynthesis.
2. Plants, animals and decomposers respire, releasing carbon dioxide.
3. Animals and plants die and are decayed by soil bacteria and fungi (**decomposers**).
4. Dead organisms can become fossilised and form the fossil fuels coal and oil. Burning fossil fuels (combustion) releases carbon dioxide into the air.

Carbon is also recycled in the sea:

1. Marine organism shells are made of carbonates. The shells drop to the sea bed as the organisms die.
2. The shells fossilise to become limestone rock.
3. Volcanic eruptions heat the limestone and release carbon dioxide into the atmosphere. Carbon dioxide is also released during weathering of the limestone rock.
4. Oceans absorbing carbon dioxide act as carbon sinks (long term store of carbon dioxide).

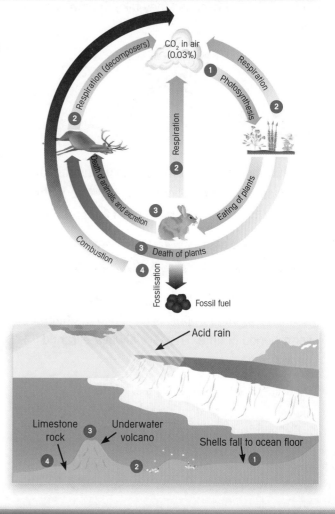

# Cycles Within Ecosystems

## The Nitrogen Cycle

The air is made up of approximately 78% **nitrogen**. Nitrogen is a vital element used in the production of **proteins**, which are needed for growth in plants and animals. Most nitrogen is stored in the air, but animals and plants can't use it because it's so **unreactive**.

The **nitrogen cycle** shows how nitrogen and its compounds are recycled in nature:

1. Plants absorb **nitrates** from the soil to make protein for growth.
2. Animals eat plants and use the nitrogen to make animal protein. Feeding passes nitrogen compounds along a food chain.
3. Dead animals and plants are broken down by decomposers, releasing nitrates back into the soil.

**Carbon** and **nitrogen** are two recycled elements.

In waterlogged or acidic soils, the recycling of nutrients takes longer. This is because waterlogged soil lacks oxygen for decomposers and acidic soil is not the best pH for decomposers.

## The Role of Bacteria

**Nitrogen-fixing bacteria** convert atmospheric nitrogen into nitrates in the soil. Some of these bacteria live in the soil, while others live in root nodules with certain plants (legumes), e.g. peas and beans.

**Nitrifying bacteria** convert ammonium compounds into nitrates in the soil.

**Denitrifying bacteria** convert nitrates and ammonium compounds into atmospheric nitrogen.

The energy released by lightning causes oxygen and nitrogen in the air to combine to form nitrogen oxides which dissolve in water. Soil bacteria and fungi act as decomposers, converting proteins and urea into ammonia.

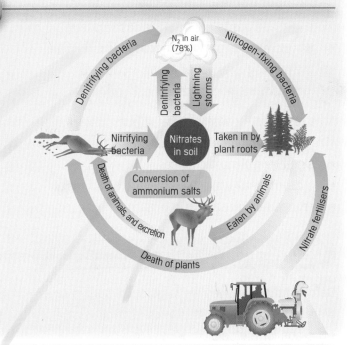

## Quick Test

1. Give the names of three natural processes that release carbon into the air.
2. Give the names of two natural processes that remove carbon from the air.
3. How much nitrogen is found naturally in the air?
4. Describe the function of nitrifying bacteria.

**Key Words**  Nitrates

# Human Influences on the Environment

## Acid Rain and Carbon Monoxide

**Acid rain** is caused by burning fossil fuels which release acid gases like sulfur dioxide and nitrogen oxides. These dissolve in rainwater to make acidic rain. This leads to metals corroding, dissolving of rocks and statues, destruction of forests, and lakes becoming acidic, killing fish and other wildlife.

**Carbon monoxide** is a poisonous gas released when we burn fossil fuels without enough oxygen. Car exhausts produce lots of carbon monoxide. Carbon monoxide combines with haemoglobin in red blood cells. Then the blood cannot carry as much oxygen as it should.

Modern cars have catalytic converters which turn carbon monoxide into carbon dioxide.

## Greenhouse Effect

Gases in the atmosphere that cause the **greenhouse effect** are: **carbon dioxide, water vapour, nitrous oxide, methane** and **CFCs**.

The greenhouse effect describes how gases, such as **methane** ($CH_4$) and carbon dioxide ($CO_2$), stop heat from 'escaping' from the Earth into Space.

More of these gases are being released due to:
- the increase in cattle and rice fields (methane)
- the burning of chopped-down wood and industrial burning (carbon dioxide).

As a result, more heat is radiated back to Earth. This is causing **global warming**, which may lead to:
- significant climate change
- a rise in sea level
- reduced biodiversity
- changes in migration patterns, e.g. in birds
- changes in the distribution of species.

Carbon dioxide can be taken from the atmosphere and stored in oceans, lakes and ponds.

Man's activities have added to the greenhouse effect, causing enhanced global warming.

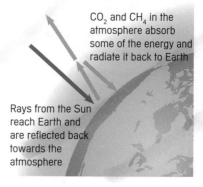

$CO_2$ and $CH_4$ in the atmosphere absorb some of the energy and radiate it back to Earth

Rays from the Sun reach Earth and are reflected back towards the atmosphere

## Deforestation

**Deforestation** involves the large-scale cutting down of trees for timber and to provide land for:
- crops for **biofuels** based on ethanol
- cattle and rice fields to provide more food (these organisms produce methane and this has led to increases in methane in the atmosphere).

Deforestation has:
- **increased** the release of carbon dioxide ($CO_2$) **into the atmosphere** through burning and the decay of wood by microorganisms
- **reduced** the rate at which carbon dioxide is removed from the atmosphere by **photosynthesis**
- **increased** the amount of **methane** in the atmosphere (produced by organisms like cattle).

Loss of forest leads to a reduction in **biodiversity**.

**Key Words**: Greenhouse effect • Global warming • Deforestation • Biodiversity

# Human Influences on the Environment

## Other Effects of Deforestation

Trees take up nutrients from soil and store them. Nutrients return to the soil when decomposers act on dead leaves. If trees are removed, the nutrients are washed away by rain. This is **leaching**.

Tree roots hold soil together. Without trees the rain can wash soil away so there is nothing for other plants to grow in. This is **soil erosion**.

When trees are cut down, transpiration is lower so the local climate is drier. This affects the **water cycle**.

Instead of being used by trees, rainwater runs straight to rivers and causes flooding.

The destruction of **peat** bogs and other areas of peat **releases carbon dioxide into the atmosphere**. Peat acts like a sponge and holds lots of rainwater. **Removal of peat** allows more rainwater to run off into rivers and **cause flooding**. Using peat-free composts could help to reduce this problem.

## Water Pollution

Animals, plants and microorganisms are all affected by water **pollution** from:
- sewage
- oil
- **fertilisers**
- pesticides
- detergents
- PCBs (chemicals used in electrical devices)

(P2) Sewage pollution results in an increase in the number of microorganisms in the water. The microorganisms use up dissolved oxygen until there is not enough for larger, active organisms.

## Eutrophication

1. **Fertilisers** or **sewage** can run into water, polluting it. They provide a lot of nitrates and phosphates, which leads to rapid growth of **algae**.
2. The algae reproduce quickly, then die and rot. They also block off sunlight, causing underwater plants to die and rot.
3. The number of **aerobic bacteria** increase and, as they feed on the dead organisms, they use up oxygen. Larger organisms then die because they can't **respire** as there is not enough available oxygen.

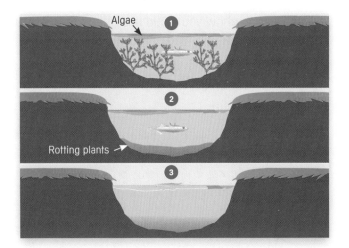

## Quick Test

1. Give one cause and one effect of acid rain.
2. Explain why carbon monoxide is harmful to humans.
3. What does the word leaching mean?
4. Why does the removal of peat cause flooding?
5. Explain why deeper water plants die if water becomes polluted with sewage.

**Key Words**  Leaching • Soil erosion • Peat • Pollution

# Exam Practice Questions

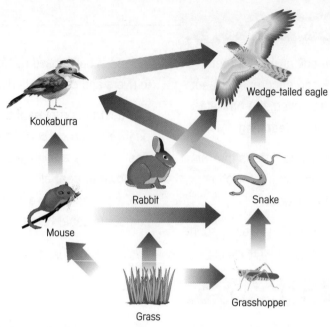

**1** From the food web above, name:

  **a)** a producer [1]

    Grass

  **b)** a secondary consumer [1]

    Mouse

  **c)** a tertiary consumer. [1]

    Eagle

**2 a)** Describe how you could use quadrats to estimate the population of grasshoppers in an area of grassland. [3]

Place quadrat, count, move on, calc average

  **b)** Grasshoppers jump about. How might this affect the reliability of your results? [2]

They may move in and out of the quadrat, tampering with the numbers grasshopper count.

# Exam Practice Questions

**3** The length of a food chain depends on the efficiency of energy transfer. Study the food chain below.

Grass (70 000kJ) → Grasshopper (12 500kJ) → Snake (2 200kJ) → Wedge-tailed eagle (120kJ)

a) Explain why this food chain does not have fourth or fifth degree consumers. [1]

*Not enough energy transfer*

b) The percentage energy transfer from producer to primary consumer is 17.9%. What is the percentage energy transfer from secondary consumer to tertiary consumer? [2]

*2200 / 120*

**4** If wedge-tailed eagles shown in the food web also eat fish from ponds, suggest how acid rain may then affect the population of snakes. [4]

*If acid rain reduces fish population, then the snake pop would also decrease as snakes are now an important food source for the eagle.*

**5** Several kinds of bacteria are important in the nitrogen cycle.

a) Name two kinds of bacteria that increase the nitrate content of the soil. [2]

b) Describe two ways in which nitrates are removed from soil. [2]

*Denitrifying bacteria*
*Plant roots*

c) Suggest how humans increase nitrates in soil. [1]

d) Excess nitrate can pollute rivers. Describe how nitrate pollution from a farm causes fish in a river to die. [4]

*Nitrates in a pond leads to Eutrophication.*

# Food Production – Crop Plants

## Artificial Controls

The rate of photosynthesis is affected by three factors:
- light intensity
- temperature
- carbon dioxide levels

The rate of photosynthesis can therefore be controlled by increasing / decreasing the levels of these factors (see page 19):

An increased rate of photosynthesis can result in plants...
- growing more quickly
- becoming bigger and stronger.

## Increasing Crop Yield

**Greenhouses** and **polythene tunnels** can be used to **control the rate** of photosynthesis, and therefore increase crop yield.

Plants grow quickly when a greenhouse is kept at the optimum temperature for the enzymes that control photosynthesis, and there is a plentiful supply of light and carbon dioxide.

**Glasshouses** and **polythene** tunnels trap heat from the Sun, and artificial lighting and heating can be used when the external conditions are dull or cool. Computer-controlled systems operate heaters, roof vents and lighting to ensure that light and temperature remain at the optimum level.

Carbon dioxide levels are increased by using paraffin heaters to supply heat. As the paraffin burns, carbon dioxide is released into the air of the glasshouse.

Growing crop plants under cover also reduces the risk of attack by pests, and diseases are easier to prevent and treat. The soil or liquid that plants are grown in is sterilised and pesticides or biological controls keep pests under control.

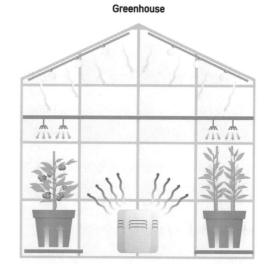

Greenhouse

Polythene Tunnel

# Food Production – Crop Plants

## Fertilisers

**Essential minerals** are needed to keep plants healthy and growing properly. Plants absorb dissolved minerals in the soil through their roots.

The minerals are **naturally present** in the soil, although usually in quite **low concentrations**. So farmers use **fertilisers** (often called NPK fertilisers) containing essential minerals to make sure that plants get all the minerals they need to grow.

Each mineral is needed for a different purpose:
- **Nitrates ($NO_3^{2-}$)** – to make proteins for cell growth.
- **Potassium compounds (K)** – for respiration and photosynthesis.
- **Phosphates (P)** – for respiration and cell growth.
- **Magnesium ($Mg^{2+}$)** – is part of chlorophyll that the plant needs for photosynthesis.

Fertilisers are expensive so the amounts used must be carefully calculated in order to obtain the maximum growth of the crop. If more fertiliser is used than is needed by the crop, then the profit gained from the crop will drop due to wasted fertiliser. Wasted fertiliser could then pollute rivers and streams.

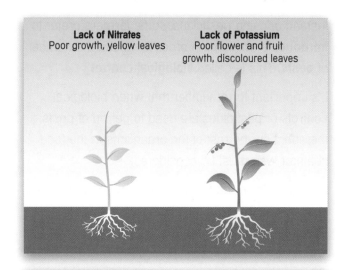

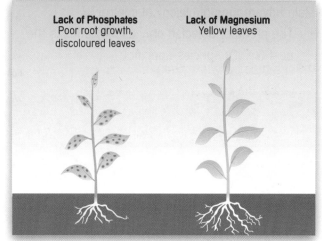

## Pest Control

Intensive farming methods try to produce as much food as possible. Farmers use **pesticides** to kill pests that **damage crops or livestock** so that **more food is produced**.

There are different types of chemical that can be used:
- **Pesticide** – kills any type of organism that harms crops or farm animals.
- **Insecticide** – a type of pesticide that kills insect pests.
- **Fungicide** – a type of pesticide that kills fungi.
- **Herbicide** – kills weeds that compete with crops for water and nutrients.

Care needs to be taken with pesticides because:
- they can harm non-pest organisms
- they can **build up (accumulate) in food chains**, harming animals at the top
- some pesticides stay **in the food chain for many years**.

**Key Words**: Pesticide

# Food Production – Crop Plants

## Biological Control

Instead of using pesticides, some farmers prefer to **introduce a predator** in order to reduce the number of **pests**. This is called **biological control**.

It's important to remember that when biological controls or pesticides are used to get rid of pests, the effect on the rest of the organisms in the food chain or web must be considered.

For example, in the food web shown below, if a pest control was to target rabbits, this would have an effect not only on the rabbits, but also on hawks and foxes (who eat rabbits).

Main **advantages** of biological control:
- The predator selected only usually attacks the pest (i.e. it's species-specific).
- Once introduced, the predator is active for many years, so repeating the treatment isn't required.
- The pest can't become resistant to the predator (unlike pesticides).
- There is no need for chemical pesticides.

Main **disadvantages** of biological control:
- The pest is reduced but it isn't completely removed.
- The predator may not eat the pest or it may even eat useful species.
- The predator may reproduce out of control and became a pest itself.
- The predator may leave the area.

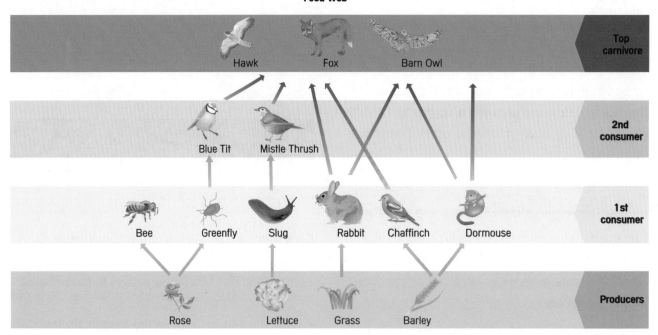

**Food Web**

## Quick Test

1. Name three factors that can limit the rate of photosynthesis.
2. What are nitrates used for in a plant?
3. Explain why lack of magnesium results in poor plant growth.
4. List three common minerals found in fertilisers.
5. What is a herbicide?
6. Explain what is meant by a biological control.

**Key Words**: Predator • Pest

# Food Production – Microorganisms

## Yeast

**Yeast** is a single-celled **fungus**. This common microorganism has been used for centuries to produce the alcohol in alcoholic drinks, and the carbon dioxide that causes bread to rise.

Yeast cells reproduce very quickly under the right conditions. They need:
- lots of sugar
- optimum temperature and pH
- the removal of waste products, such as **alcohol**, which poison the yeast.

When yeast reproduces, the chromosomes are copied and a new nucleus is made. The new cell 'buds' off the parent. This is known as 'budding'.

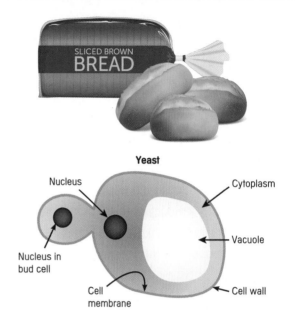

### Temperature and Yeast Growth

The growth rate of yeast doubles with every 10°C rise in temperature. So, increasing the temperature increases the rate of **growth**. But, above 40°C, the yeast enzymes are **denatured**, which causes the growth rate to slow down and stop.

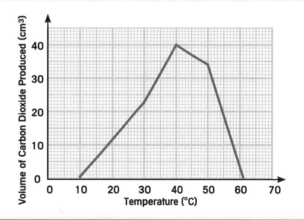

## Fermentation

**Fermentation** is **anaerobic respiration** in yeast. It produces alcohol and is used to make alcoholic drinks.

Sugars are broken down by yeast in the absence of oxygen to produce the alcohol.

Different fruits and seeds are used to provide the **yeast** with sugars and to give a drink flavour.

**Carbon dioxide** is also produced during fermentation:

| glucose (sugar) → ethanol (alcohol) + carbon dioxide |
| --- |
| $C_6H_{12}O_6 \rightarrow 2C_2H_5OH + 2CO_2$ |

Yeast cells feed on **sugars**. They can respire with oxygen (**aerobic respiration**) or without oxygen (**anaerobic respiration**) to release energy from sugar.

Brewers obviously want yeast to carry out anaerobic respiration as alcohol is a by-product.

**Key Words**: Yeast • Fermentation

# Food Production – Microorganisms

## Brewing Bear

This is how beer is brewed:

1. Extracting sugar – barley seeds are mixed with water and allowed to sprout, turning the starch in the seeds into sugars.
2. Hops are added to give flavour to beer.
3. Yeast is added to ferment the sugars into alcohol. The mixture is kept warm so the yeast **reproduces** and **respires**.
4. The tank is sealed so the yeast can respire **anaerobically** producing alcohol. This also stops unwanted microorganisms spoiling the beer.
5. A chemical is added to make the yeast settle, leaving a clear liquid. This is called **clarifying** or **clearing**.
6. The beer is **pasteurised** by heating it to 72°C for 15 seconds. This kills harmful microorganisms but doesn't greatly affect the taste.
7. The beer is bottled or put in sealed casks (bottling or casking).

Yeast must be filtered out (or killed by heat treatment) if the beer is going to be bottled. Otherwise it would continue to respire, producing carbon dioxide that would make the bottles explode.

## Investigating Respiration Rate in Yeast

The apparatus shown here can be used to compare the rate at which yeast gives off carbon dioxide in different conditions.

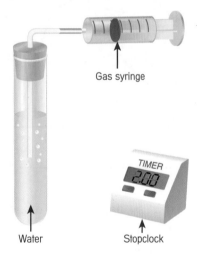

In this experiment you **count the number of bubbles given off over a period of time**, e.g. 1 minute, 10 minutes, etc. The experiment can be repeated at a range of **different temperatures** by standing the flask in a water bath.

The effect of **varying the concentration of glucose** on the rate of respiration can also be tested by setting up flasks with a **range of glucose concentrations**. The same apparatus can be used to investigate the rate of respiration in **different varieties of yeast**.

### Variables

**The dependent variable** is the rate of respiration, indicated by the number of bubbles of carbon dioxide per minute. **Independent variables** that can be tested are: **temperature, concentration of glucose, type of sugar,** and **variety of yeast**. Other variables that should be kept **constant** are: **volume of solution, mass of yeast used, time that you counted from start of experiment,** and **diameter of the delivery tube**.

### Control

A **control flask** can be set up, which contains **boiled yeast that is then cooled, instead of using living yeast**. The control should be tested in the same conditions as the experiment tubes. Because the yeast is dead you can use it to make sure **that any gas produced in the experiment is from the respiration of the yeast, not from any other chemical reaction** happening in the tube.

### Improvements

The gas given off can be collected in a gas syringe to measure the volume of carbon dioxide produced per minute, as a more accurate way of investigating respiration rate.

**Key Words** — Pasteurise

# Food Production – Microorganisms

## Making Yoghurt

Yoghurt is made in large steel **fermenters**:

1. The equipment is sterilised using steam to kill any **pathogens**.
2. Raw milk is heated to 80°C to kill bacteria, and then quickly cooled. This is **pasteurisation**.
3. A live bacterial (*Lactobacillus*) culture is added to the warm milk. The mixture is **incubated for several hours**.
4. The bacteria reproduce and **feed on the lactose** sugar in the milk, producing **lactic acid**, which gives a sharp taste to the yoghurt and thickens and preserves it.
5. The manufacturer samples the yoghurt for consistency and flavour.

Flavours and colours might be added before packaging.

## Industrial Fermenters

Microorganisms can be grown in **industrial fermentation tanks**. The organisms are mixed with **nutrient medium** and are kept at the **optimum temperature and pH** so that they grow as fast as possible. They are provided with **oxygen** so that they can respire. The conditions inside the fermenter are therefore ideal for the growth of many microorganisms.

Before each batch of organisms is grown, the fermenter must be thoroughly cleaned and **sterilised**. All nutrients put into the fermenter must be sterile. The oxygen passes through a **filter to remove microorganisms** before it reaches the tank.

Only the organism you want to grow is allowed into the fermenter. Otherwise, unwanted organisms could easily multiply. This would waste the nutrients supplied and be expensive. Some of the organisms could be harmful.

Microorganisms are grown on a large scale to make many things, e.g. **food, enzymes, antibiotics** and **medicines like insulin**.

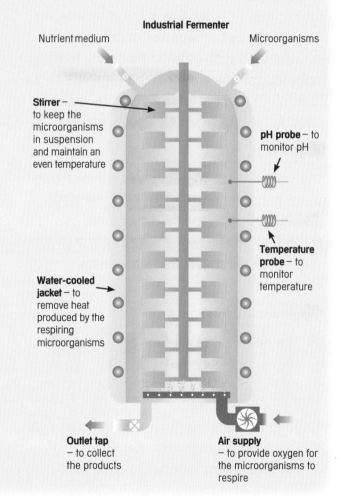

## Quick Test

1. Which kind of respiration produces alcohol?
2. *P2* Name the bacterium used to make yoghurt.
3. *P2* Describe how bacteria are used to make yoghurt.
4. *P2* Name four useful products of industrial fermentation.
5. *P2* Explain why fermenters must be sterilised.

# Food Production – Fish

## Fish Farming

Fish are an **economical source of protein** because they are cold blooded and so don't need to use energy to keep warm – instead they use it to grow.

Fish stocks in the sea are falling due to overfishing, so some fish are now 'farmed'. Fish farming **produces less carbon dioxide and methane** (greenhouse gases) than farming mammals.

### Tanks and Cages

Freshwater fish are farmed in tanks on land, saltwater fish in cages in the sea. **Fish cages** allow seawater to flow through, bringing **well-oxygenated water** and carrying waste products away. **Freshwater tanks** are oxygenated and have a continuous flow of water. The water flowing through the tanks must be free of harmful chemicals and contain enough oxygen.

### Cleaning

The tanks must be **regularly cleaned** to remove waste products. Waste products increase the growth of microorganisms, which could harm the fish and spread disease. If waste from the fish escapes, **it can pollute the surrounding ecosystem**.

### Increasing Yield

Fish are kept at high densities.
- High-quality food is used.
- Antibiotics and pesticides are used to protect the fish from disease.
- Tanks and containers are kept free from fish waste.
- Fish are protected from predators.
- Selective breeding is used, i.e. good quality fish are chosen to breed.

### Fish Pests and Diseases

Farmed fish are kept **intensively**. The high density of fish in the cages and tanks causes stress. Stress makes fish more likely to suffer from infection. Farmers treat the fish in the cages and tanks with **antibiotics and pesticides**.

**Hazard:**
- In some systems it is difficult to prevent these **chemicals from polluting surrounding ecosystems**. Some of the antibiotics and **pesticides accumulate** in the fish, which we then eat.

### Feeding

Fish that are herbivores are kept in ponds fertilised with nutrients to encourage growth of pondweed as food.

Seawater fish are mostly carnivorous and are fed on fish meal. So, wild fish still have to be caught to make fish meal – 4kg of wild fish are needed to feed 1kg of farmed fish.

**Hazards:**
- Fertilised water must be kept within the pond to **prevent eutrophication** in surrounding lakes / rivers.
- There is still a danger of **overfishing wild fish** in order to provide food for seawater fish.

### Predation

Fish farms attract predators, e.g. birds, dolphins, seals and otters. This is **interspecific predation**. Strong **netted cages** protect the fish.

The **density** of fish in the tanks is monitored and **kept as low as is profitable**, so that fish don't eat each other. Fish eating others in the same population is **intraspecific predation**.

## Quick Test

1. Why are farmed fish prone to stress?
2. How do fish farmers protect their stock from pests and diseases?
3. Suggest three ways in which fish farming can harm the environment.
4. What is intraspecific predation?

**Key Words** — Interspecific predation • Intraspecific predation

# Selective Breeding

## Selective Breeding

**Natural selection** and **selective breeding** are very similar, but one is natural and the other is carried out by humans for their own purposes, not to suit the environment / help the survival of species.

**Selective breeding** is when plants or animals with certain traits are reproduced to produce offspring with certain desirable characteristics. Selective breeding can produce two outcomes:

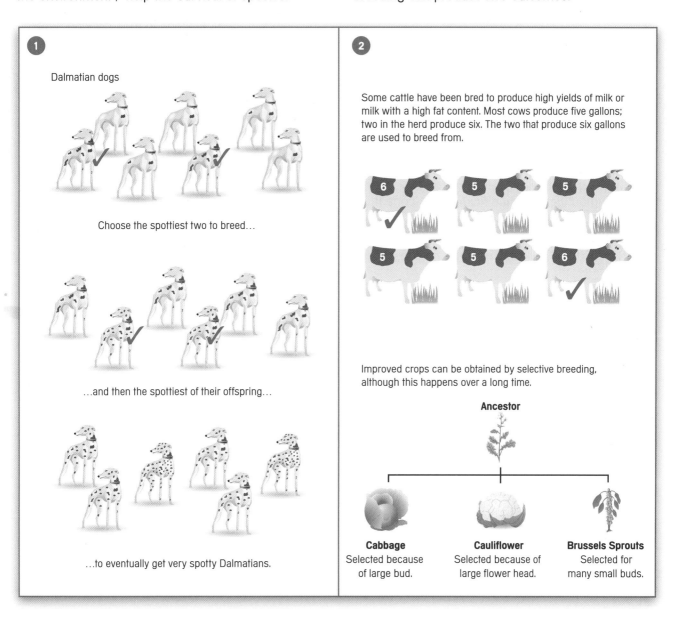

**1**

Dalmatian dogs

Choose the spottiest two to breed…

…and then the spottiest of their offspring…

…to eventually get very spotty Dalmatians.

**2**

Some cattle have been bred to produce high yields of milk or milk with a high fat content. Most cows produce five gallons; two in the herd produce six. The two that produce six gallons are used to breed from.

Improved crops can be obtained by selective breeding, although this happens over a long time.

**Ancestor**

**Cabbage**
Selected because of large bud.

**Cauliflower**
Selected because of large flower head.

**Brussels Sprouts**
Selected for many small buds.

## Quick Test

1. Explain how lettuces with curly leaves could be produced through selective breeding.

**Key Words**  Selective breeding

# Genetic Modification (Genetic Engineering)

## Uses of Genetic Engineering

Genetic engineering can be used to:
- make vaccines
- make medicines
- make useful human proteins like insulin and human growth hormone
- improve crop plants.

(P2) Animals and plants whose cells have been modified by the transfer of genetic material from a different species are called **transgenic.**

Genes can also be transferred to the cells of animals and plants at an early stage in their development so that they develop with **desired characteristics**.

## Transferring Genes

DNA contains the code for the protein a particular organism needs. Proteins produced by one organism may not be produced by another.

By carrying out **genetic modification**, the gene that produces a desirable protein can be inserted into another organism so that it too produces the right protein.

Different types of enzyme are used in genetic engineering:
- **Restriction enzymes** are used to cut open DNA leaving 'sticky' ends.
- **Ligase enzymes** are used to rejoin DNA strands. The 'sticky' ends rejoin the DNA strands.

## Using Vectors in Genetic Engineering

**Vectors** are used in genetic engineering to replicate useful genes and transfer them to the cells that scientists want to modify.

Plasmids are small loops of DNA found in the cytoplasm of bacteria. Because the plasmid loops can be taken up by bacteria, they can be used as **vectors** in genetic engineering.

In some processes genes are transferred from the donor cell to the new cell by viruses. Such viruses are acting as gene vectors too.

## Genetic Modification and Food Production

Crops that have had their genes modified are called **GM crops**. Crop plants can be genetically modified to be resistant to pests.

**Advantages** of GM crops:
- The resistant plants will be able to grow more rapidly and produce a greater yield (more food per plant) because they are not damaged by pests. Food production will increase.
- Costs will decrease because farmers don't need to spray crops with pesticides.

**Disadvantages** of GM crops:
- There are fears that the gene for resistance will spread to plants other than the crop through **cross-pollination**. This would mean wild flowers would also become pest resistant and so other organisms further down the food chains that depend on the pests would be affected.
- There are fears that introducing new genes into food may harm humans, although so far research suggests this is unlikely.

**Key Words**: Genetic modification • Restriction enzyme • Ligase enzyme • Vector

# Genetic Modification (Genetic Engineering)

## Producing Human Insulin

Human **insulin** is produced using the following method:

1. The human gene for insulin production is identified. It's removed using a special **restriction enzyme** which **cuts through the DNA** in precise places.
2. The enzyme is then used to **cut open a loop of bacterial DNA** (plasmid) in the cytoplasm.
3. Ligase enzymes are used to **insert the section of human insulin gene** into the plasmid.
4. The plasmid is **taken up by bacteria** which start to divide rapidly. As they divide, they replicate the plasmids that now contain the human gene to make insulin.
5. The **transgenic** bacteria are cultured by cloning in large **fermenters**. Each bacterium carries instructions to make insulin. The cultured bacteria are allowed to make the protein and commercial (large) quantities of insulin can then be harvested and purified.

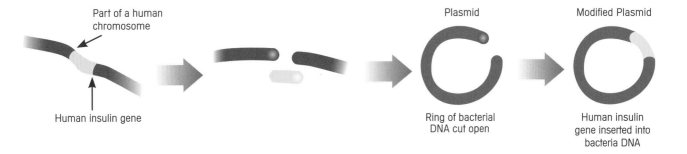

## Benefits and Concerns about GM Insulin

**Benefits:**
- **Fewer adverse reactions** – for many years insulin was extracted from pigs, and diabetics who used it suffered side effects due to immune reactions caused by the difference between human insulin and pig insulin. Human insulin that is produced using genetically modified bacteria produces fewer side effects.
- **Cost** – human insulin can be mass produced economically using genetically modified organisms.

**Concerns:**
- **Antibiotic resistance** – there is concern that the process of modifying bacteria uses antibiotic resistant genes, which are replicated in the fermenters along with the desired genes. Releasing more bacteria that carry antibiotic resistance genes into the environment may make it even more difficult to combat antibiotic resistance in general.

## Quick Test

1. Name two enzymes used in genetic engineering and state what they do.
2. What is a plasmid?
3. How are vectors used in genetic engineering?
4. What are fermenters used for in the production of human insulin?
5. Suggest one reason why genetically modifying crop plants are a good idea.

# Cloning

## Cloning

**Cloning** is an example of asexual reproduction which produces genetically identical copies. Identical twins are **naturally occurring** clones.

**Animals** and plants can be **cloned artificially**. The most famous example is Dolly the sheep, who was the first mammal to be successfully cloned from an adult body cell.

### Adult Cell Cloning – Plants

Plants can be cloned to be sold commercially.

| Advantages |
|---|
| • The cloned plants will be genetically identical to the parent, so all the characteristics will be known. |
| • It is possible to mass produce plants that may be difficult to grow from seeds. |
| **Disadvantages** |
| • Any susceptibility to disease or sensitivity to environmental conditions will affect all the plants. |
| • The reduction in **genetic variation** reduces the potential for further selective breeding. |

### Cloning by Tissue Culture

Cloned plants can be produced by the following method:
1. Select a parent plant with desired characteristics.
2. Scrape off a lot of small pieces of tissue into beakers containing nutrients and hormones. Make sure that this process is done **aseptically** (without the presence of bacteria or fungi) to avoid the new plants rotting.
3. Lots of genetically identical plantlets will then grow (these can also be cloned).

Many older **plants** are still able to **differentiate** or **specialise**, whereas animal cells lose this ability. So cloning plants is easier than cloning animals.

### Adult Cell Cloning – Animals

The following method was used to produce a cloned sheep (i.e. Dolly):

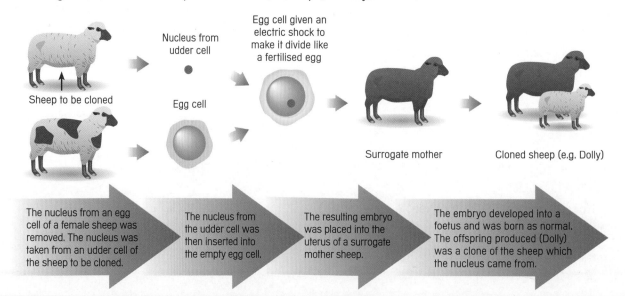

The nucleus from an egg cell of a female sheep was removed. The nucleus was taken from an udder cell of the sheep to be cloned.

The nucleus from the udder cell was then inserted into the empty egg cell.

The resulting embryo was placed into the uterus of a surrogate mother sheep.

The embryo developed into a foetus and was born as normal. The offspring produced (Dolly) was a clone of the sheep which the nucleus came from.

# Cloning

## Uses of Cloning

There are a number of uses of cloning:
- It's possible to clone human embryos in the same way that animals are cloned. This technique could be used to provide **stem cells** for medical purposes.
- The mass production of animals with desirable characteristics.
- Producing animals that have been genetically engineered to provide human products.

There are major ethical dilemmas about cloning humans:

- The cloning process is very unreliable – the majority of cloned embryos don't survive.
- Cloned animals seem to have a limited life span and die early.
- The effect of cloning on a human's mental and emotional development isn't known.
- Some religions say that cloning humans is wrong.
- For some people, using human embryos and tampering with them is controversial.

Cloning whole humans is illegal, but tissue cloning is legal.

## Benefits and Risks of Animal Cloning

There are benefits and risks associated with cloning technology.

**Benefits** of cloning:
- Genetically identical cloned animals will all have the same characteristics.
- The sex of an animal and timing of birth can be controlled.
- Top-quality bulls and cows can be kept for egg and sperm donation, whilst other animals can be used to carry and give birth to the young.

**Risks** of cloning:
- Cloning reduces genetic variation.
- Cloned animals are identical copies, so they are all genetically the same. There is potential for one disease being able to wipe them all out.
- Welfare concerns – cloned animals may not be as healthy or live as long as 'normal' animals.

## Animal Organ Donors

There is a **shortage** of **human organ donors** for **transplants**.

One possible solution would be to **genetically engineer** (i.e. artificially alter the genetic 'code of) an **animal** so its organs wouldn't be rejected by the human body. The animal could then be **cloned** to produce a ready supply of identical donor organs as well as antibodies.

Animal organ donors could solve the problem of waiting lists for human transplants.

However, there are:
- **concerns** that infections might be passed from animals to humans
- **ethical issues** concerning animal welfare and rights.

## Quick Test

1. Why is it easier to clone plants than animals?
2. Give two advantages of cloning plants.
3. Suggest two reasons why some people disagree with cloning animals.

**Key Words**    Stem cell

# Exam Practice Questions

**1** The diagram shows an industrial fermenter used to produce the antibiotic penicillin. The microorganism *Penicillium chrysogenum* is grown in the nutrient liquid. The organism secretes the antibiotic penicillin into the liquid.

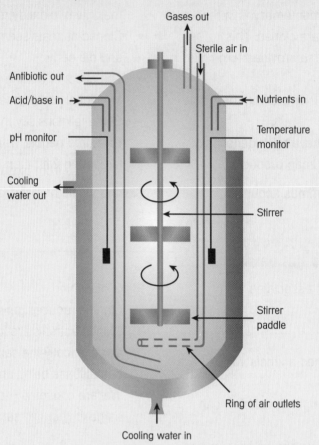

Explain why the following parts of the fermenter are needed:

**a)** the stirrer [2]

*Mixes the solution, and therefore speeding up the reaction as a whole.*

**b)** the sterile air [2]

**c)** nutrient medium [3]

# Exam Practice Questions

**d)** temperature monitor [3]

**2** Human insulin is produced by genetically modified bacteria that are grown in a fermenter. The human gene for insulin is added to the bacteria, which then grows and secretes insulin. The insulin is then separated from the contents of the fermenter. Insulin was produced from pigs before genetic modification was developed.

**a)** Describe how the gene for insulin is transferred to the bacteria. [5]

**b)** Give two reasons why it is important that the contents of the fermenter are sterile before the bacteria are added, and that anything entering the fermenter is also sterile. [2]

Reason 1:

Reason 2:

**c)** Suggest two advantages of producing human insulin through genetic engineering. [2]

# Exam Practice Questions

**3** Crop plants can be genetically modified to be resistant to pests.

   **a)** Suggest two reasons why farmers want pest-resistant crops. [2]

   .................................................................................................................................................................

   .................................................................................................................................................................

   .................................................................................................................................................................

   **b)** Give two reasons why some people are against the production of genetically modified crops. [2]

   .................................................................................................................................................................

   .................................................................................................................................................................

   .................................................................................................................................................................

**4** Yeast is a single celled organism that reproduces very fast in optimum conditions. It is used in the brewing of beer. When yeast is mixed with water it makes a cloudy solution.

A student investigated how fast yeast grows in different temperatures. He set up flasks of yeast and glucose solution at different temperatures. He covered each solution with a layer of oil. Every 30 minutes for two hours he held a card behind each flask to see if he could see the X on it through the solution.

| Temperature (°C) | Time at which X can no longer be seen (mins) |
|---|---|
| 20 | 90 |
| 30 | 60 |
| 40 | 30 |
| 50 | |

The student stopped the experiment after 2 hours. The X in the tube at 50°C was still visible.

   **a)** Explain why the student added glucose. [3]

   .................................................................................................................................................................

   .................................................................................................................................................................

   **b)** Which is the dependent variable in this experiment? [1]

   .................................................................................................................................................................

98

# Exam Practice Questions

**4 c)** State two variables the student should control to make the test fair. **[2]**

**d)** Explain why the X disappeared in three out of the four tests. **[2]**

**e)** At which temperature did the yeast grow fastest? **[1]**

**f)** Suggest why there is no result for 50°C. **[2]**

**g)** Oil poured on the surface stops oxygen reaching the yeast. What kind of respiration will the yeast use? **[1]**

**h)** Name a substance produced during this process that is useful in brewing. **[1]**

**i)** Suggest how the student could improve the experiment. **[3]**

# Answers

In the examination, there are three assessment objectives that are tested throughout the exam papers:

**AO1 Knowledge and understanding** (worth approx. 45–50% of marks in the exam)
**AO2 Application of knowledge and understanding, analysis and evaluation** (worth approx. 27–33% of marks in the exam)
**AO3 Experimental skills, analysis and evaluation of data and methods** (worth approx. 20–25% of marks in the exam)

## The Nature and Variety of Living Organisms

### Answers to Quick Test Questions

**Page 4**
1. Respiration
2. Organisms grow by increasing the number and size of their cells.
3. Organisms regulate their water content by controlling the amount of water they take in and the amount of water they get rid of.
4. Plants respond to a change in the direction of light.
5. Organelles move through the cytoplasm of cells. Ciliated cells waft their cilia. (Other suggestions that you could make include cells that use flagellae to move the whole cell and cells that move along by using pseudopodia.)

**Page 7**
1. Protoctista, bacteria, fungi, plants, animals
2. Plants store carbohydrates as sucrose or starch; animals store carbohydrates as glycogen.
3. The tobacco mosaic virus stops the plant making chloroplasts. This means that the plant cannot photosynthesise and so will not be able to grow well.
4. HIV destroys cells of the immune system. When a person with HIV is infected by the cold virus their immune system is not able to destroy the pathogen easily and so they become very ill.

### Answers to Exam Practice Questions

AO2 1

| Animals | D |
|---|---|
| Plants | B |
| Bacteria | E |
| Fungi | C |
| Protoctists | A |

[1 mark for each correct answer]

AO1 2. Plant cell walls are made of cellulose whereas cell walls of fungi are made of chitin **[1 mark]**; Some plant cells contain chloroplasts whereas cells of fungi do not **[1 mark]**.

In questions like this, be careful to distinguish between structural and functional differences. For example, **you would gain no marks** in this question for explaining that some plant cells can photosynthesise, whereas animal cells cannot.

In the same way you might be asked to list structural differences between arteries and veins. **No marks are available** for saying that arteries carry blood away from the heart whereas veins carry it towards the heart, even though this is true.

Can you think of one **structural** difference between sensory and motor neurones, and also one **difference in function** between them?

AO1 3. a) Hyphae b) Mycelium
AO1 4. Fungi secrete digestive enzymes onto the food and large food molecules are broken down into small molecules **[1 mark]** by extra-cellular digestion **[1 mark]**. The products of digestion are absorbed into the cells of the fungus **[1 mark]**.

AO1 5.

| Group | Organism | Disease |
|---|---|---|
| Bacteria | Pneumococcus | Pneumonia |
| Virus | HIV | AIDS |
| Protoctista | *Plasmodium* | Malaria |

[1 mark for each correct answer]

AO1 6. Rod-shaped **[1 mark]**; bacterium **[1 mark]**; used to make yoghurt **[1 mark]**

AO1 7. a) [1 mark for each correct label]

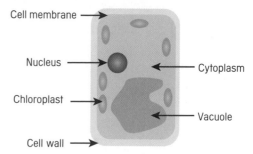

AO1 b) Chloroplast **[1 mark]**; vacuole **[1 mark]**; cell wall **[1 mark]**

## Structures and Functions in Living Organisms

### Answers to Quick Test Questions

**Page 12**
1. Mitochondria
2. A cell wall, chloroplasts and a permanent vacuole.
3. A tissue
4. An organ
5. Epithelial tissue

**Page 14**
1. Carbon, hydrogen, oxygen and nitrogen
2. a) Sugars b) Fatty acids and glycerol c) Amino acids
3. Mix a sample of the cereal with water in a test tube. Add a drop of Benedict's Reagent and then heat the mixture carefully. If the cereal contains glucose, the Benedict's Reagent will turn green, orange or red.
4. Enzymes are biological catalysts made of protein. They speed up metabolic reactions.
5. Enzymes are highly specific. Only one shape of substrate will fit into any particular active site.
6. As temperature increases, both the enzyme and the substrate move more quickly. This increases the chances of the substrate colliding with the active site. The more frequent the collisions, the faster the reaction will go.

# Answers

7. High temperatures and extreme pH change the shape of the enzyme molecule and alter the shape of the active site. The substrate no longer fits and so the enzyme cannot speed up the rate of the reaction. The change is irreversible.

## Page 18
1. Diffusion involves the net movement of particles from high to low concentration. Active transport involves the net movement of particles from low to high concentration. This requires energy from the cell and is therefore an active process.
2. **Any three from:** Steep concentration gradient; Large surface area across which to diffuse; Increasing the temperature; A short distance for the particles to diffuse across.
3. Osmosis is the diffusion of water from a dilute solution to a more concentrated solution, through a semi- or partially permeable membrane.
4. Water enters the plant cells by osmosis and pushes against the inelastic cell wall. The pressure inside the cells increases and holds the plant upright.
5. The seawater around the roots is a higher concentration than the contents of the root cells, so water leaves the plant by osmosis. The cells then become flaccid and the plant wilts.
6. The carrot will increase in length because its cells take in water by osmosis.

## Page 20
1. $6CO_2 + 6H_2O \longrightarrow C_6H_{12}O_6 + 6O_2$
2. When the temperature rises, the rate of photosynthesis increases. This is because the enzymes and substrates used in photosynthesis are colliding more frequently. Temperature increases above the optimum will denature the enzymes and so the rate of photosynthesis will then fall to zero.
3. Gases diffuse through the stomata, into and out of the air spaces of the leaf.
4. The cells of the spongy mesophyll are very loosely packed, with many air spaces between them. This exposes a very large surface area of cells to the air and so produces a high surface area to volume ratio.
5. The broad shape gives the leaf a large surface area exposed to sunlight. The thin shape means there is only a short distance for gases to diffuse in and out of the stomata.

## Page 22
1. Magnesium, Nitrates
2. **a)** Nitrates **b)** Magnesium
3. **a)** A leaf that has some parts that are white because they don't contain chlorophyll and other parts that have chlorophyll in them **b)** Variegated leaves only contain starch in the parts that contain chlorophyll. This shows that chlorophyll is needed for photosynthesis.
4. Trap pondweed under an inverted glass funnel in a beaker of water and illuminate the weed. Collect the bubbles of gas given off in an inverted test tube of water, then test the gas with a glowing splint. The splint relights, which shows it is oxygen.

## Page 24
1. Carbohydrates, lipids, protein, vitamins, minerals, water, dietary fibre
2. **a)** Meat and fish **b) Accept two suitable answers**, e.g. pasta, rice, cake, bread, potatoes
3. The extra energy is stored in the body as fat and the person may become obese.
4. **a)** Liver **b)** Salivary glands, pancreas, small intestine **c)** Stomach

## Page 26
1. To break the large food molecules down into molecules that are small enough to be absorbed
2. Chewing using teeth, churning using the muscles of the stomach and the emulsification of fats by bile; Mechanical digestion breaks large lumps of food into smaller ones that have a large surface area for the action of enzymes.
3. Ingestion is what occurs when food or drink is put into the mouth. Egestion is when undigested food and excess water leave the alimentary canal through the anus.
4. Description: A villus is a microscopic finger-like projection that contains capillaries. Villi project from cells lining the small intestine. Adaptation: Villi increase the surface area through which nutrients are absorbed. They are well supplied with blood, which takes absorbed nutrients away and therefore maintains a steep diffusion gradient for absorbing nutrients easily. The blood capillary is very close to the surface of each villus so the distance nutrients have to diffuse is very short.

## Page 29
1. Movement, synthesis of large molecules, active transport
2. The release of energy stored in food chemicals in all living cells
3. Aerobic respiration requires oxygen to completely break down glucose. Anaerobic respiration occurs without oxygen and releases less energy from the glucose.
4. Living peas are respiring so they release energy, some of which is in the form of heat.

## Page 31
1. **a)** All the time **b)** When the plant is in the light
2. Microscopic pores on the underside of leaves that allow gases to pass in and out
3. Air spaces allow gases to circulate inside the leaf.
4. Plants photosynthesise fast during the day and therefore produce more oxygen than they use for respiration. Plants continue to respire at night but do not photosynthesise, so they produce carbon dioxide but do not produce oxygen.

## Page 33
1. Nose, trachea, bronchi, bronchioles, alveoli
2. The cartilage rings keep the airways open.
3. Oxygen diffuses across the wall of the alveolus and through the capillary wall into the blood. Carbon dioxide diffuses from the blood into the air in the alveolus.
4. Diaphragm and intercostal muscles
5. **a)** Breathing in: The diaphragm and intercostal muscles contract, so the ribs rise and the diaphragm flattens. The pressure inside the thorax falls and so air enters the lungs.
   **b)** Breathing out: The diaphragm and intercostal muscles relax, so the ribs move down and the diaphragm moves up in the thorax. This increases the pressure inside the thorax, which pushes air out of the lungs.
6. Some cells lining the trachea and bronchi secrete mucus. Most of the lining cells have cilia. The particles stick in the mucus. The mucus is moved up out of the lungs by the waving action of the cilia.

## Page 35
1. Heart disease, cancer, bronchitis, emphysema
2. Tar, carbon monoxide, nicotine
3. Carbon monoxide is picked up by the blood instead of oxygen. The body then doesn't get sufficient oxygen, so breathes faster to try to get more.
4. During exercise, the muscle cells are respiring faster. They need more oxygen to do this and produce more carbon dioxide as a result. The heart rate increases to speed up the flow of blood and deliver oxygen faster, and to remove the extra carbon dioxide more efficiently.

## Page 37
1. Transpiration is the loss of water vapour from a leaf, by evaporation from cell surfaces and diffusion through the stomata.
2. **a)** Water and mineral salts **b)** Amino acids and sucrose
3. **Any three from:** Higher temperature; Wind; Low humidity; Increased light intensity
4. Water enters root hair cells from the soil by osmosis across the partially permeable cell membrane.
5. Use a mass potometer. Submerge the roots of a leafy plant in water in a sealed bag. Put the bag in a beaker and stand it on a digital balance. Record the fall in mass due to transpiration.

# Answers

## Page 39
1. To measure the rate of transpiration
2. It causes the stomata to open.
3. Higher temperatures increase transpiration rate because they increase the rate of evaporation of water from cells inside the leaf. This increases the concentration gradient for water vapour between the leaf air spaces and the atmosphere, so water vapour diffuses more quickly out through the stomata.
4. High humidity decreases transpiration because the high concentration of water vapour outside the stomata reduces the concentration gradient between the leaf air spaces and the surrounding air. So diffusion of water vapour out of the leaf is slower.
5. Hot, sunny and dry conditions

## Page 41
1. Platelets, plasma, red blood cells, white blood cells
2. **Any three from:** Small and flexible so can bend through narrow capillaries; Large surface area to volume ratio to absorb oxygen efficiently; Contain haemoglobin which combines easily with oxygen; Do not have nucleus and so can carry a lot of haemoglobin
3. Arteries have thick muscular and elastic walls, which can withstand the high pressure of blood they carry. Veins carry blood at low pressure. They contain valves that prevent the low pressure blood flowing backwards. They also have thinner walls than arteries as they don't need to withstand high pressures of blood.

## Page 44
1. The body provides ideal conditions for microorganisms to grow – warmth, moisture and plenty of nutrients.
2. Phagocytes and lymphocytes
3. Some antibodies make microorganisms clump together and this makes it easier for phagocytes to digest them.
4. A vaccination causes production of memory cells to a specific disease-causing organism. If a person is later infected with that organism the memory cells stimulate very rapid production of antibodies to the organism. The organism is destroyed before it makes the person ill.

## Page 47
1. The removal, from an organism, of the waste products from chemical reactions in its cells.
2. Oxygen, from photosynthesis; carbon dioxide, from respiration
3. Glomerulus, Bowman's capsule, proximal convoluted tubule, loop of Henlé, distal convoluted tubule, collecting duct, ureter, bladder
4. In a healthy kidney all the glucose that is filtered out of the glomerulus is reabsorbed in the proximal convoluted tubule.
5. ADH (anti-diuretic hormone)
6. Small volume of dark-coloured, concentrated urine produced

## Page 49
1. The maintenance of constant internal conditions in the body
2. Temperature, concentration of the internal environment
3. Blood vessels in the skin dilate so more heat can be transferred to the environment. We begin to sweat and the sweat uses heat energy from the body to evaporate.
4. Liver; Insulin reduces blood glucose concentration by making the liver cells convert glucose into glycogen
5. Oestrogen and progesterone

## Page 50
1. a) Roots b) Roots c) Shoots
2. Auxin in the shoot makes cells grow longer. Light causes auxin to accumulate on the shaded side of the shoot, so the shaded cells grow longer than those on the side near the light. This makes the shoot bend towards the light.
3. Positive phototropism results in plant leaves getting more light, so they photosynthesise more efficiently.

## Page 51
1. **Any two from:** Nerves – electrical impulses, hormones – chemical; Nerves to effectors, hormones to target organs; Impulses in nerves, hormones via blood; Impulses travel fast, hormones slower; Impulses short acting, hormones long-term effects; Impulses precise area of response, hormones affect whole body.
2. Motor neurone

## Page 53
1. Reflex actions protect the body from harm because the response occurs much more rapidly than if it was a conscious action.
2. Retina
3. Cornea and lens
4. Ciliary muscles relax which tightens the suspensory ligaments.
5. Radial muscles of the iris

### Answers to Exam Practice Questions

**AO1** 1. a) A: Loop of Henlé **[1 mark]**; B: Glomerulus **[1 mark]**; C: Collecting duct **[1 mark]**; D: Bowman's capsule **[1 mark]**; E: Proximal convoluted tubule **[1 mark]**

**AO1** b) i) B and D **[1 mark for each correct answer]**
ii) Water is reabsorbed (osmoregulation)

**AO3** c) 0g/day

**AO2** d) Blood proteins are too large to pass out of the glomerulus into the Bowman's capsule.

**AO2** e) Glucose molecules are small enough to be filtered into the Bowman's capsule **[1 mark]**; all the glucose molecules are reabsorbed into the blood **[1 mark]**; by diffusion and active transport **[1 mark]**

**AO2** 2. a) Heat a urine sample with Benedict's Reagent **[1 mark]**; a change in colour to red/orange indicates that glucose is present **[1 mark]**

**AO2** b) The high pressure damages the delicate glomerulus walls **[1 mark]** and allows large protein molecules to be forced through **[1 mark]**.

**AO1** 3. a) The loss of water vapour **[1 mark]**; by evaporation and diffusion from leaves **[1 mark]**

**AO3** b) i) 120g
ii) $\frac{120}{4}$ **[1 mark]** = 30g/hour **[1 mark]**

> You will often receive a mark for a calculation, even if the answer is incorrect. So it is very important to show all your working.

**AO3** c) x-axis time/hours OR time/mins AND y-axis mass/g, AND values given for each large grid square on both axes **[1 mark]**; scale easy to plot and graph more than half size of grid **[1 mark]**; plotted points accurate within one mm **[maximum of 2 marks: delete 1 mark per error]**; points joined accurately with ruled lines **[1 mark]**

**AO3** d) Water lost by evaporation from leaves / transpiration causes loss in mass

**AO2** e) Steady loss in mass for first 60 minutes **[1 mark]**; more rapid loss in mass between 60 and 240 minutes **[1 mark]**.

> **Describe** and **Explain** mean different things! You will often be asked to describe the changes that the graph shows and students often lose marks because they write why the changes are happening instead of describing the shape of the line they see (in other words they explain the graph).
>
> When you **describe** a graph you should always try to refer to the figures on the axes as well.

**AO3** f) Higher light intensity when Sun out **[1 mark]**; causes stomata to open so transpiration increases **[1 mark]**; increase in temperature after one hour **[1 mark]**; increases rate of evaporation (so transpiration increases) **[1 mark]**.

# Answers

AO1 **4. a)** Carbon dioxide and oxygen

> Be careful with lists. If you are asked for two things, make sure you give just two, not three things. Sometimes you can lose marks if you give an extra answer that is wrong.

AO1 **b)** Carbon dioxide is produced from respiration **[1 mark]**; oxygen is produced from photosynthesis **[1 mark]**.

AO2 **c)** Thin flat shape / air spaces between mesophyll cells **[1 mark]**; provides large surface area to volume ratio (for gas exchange) **[1 mark]**; loose arrangement of spongy mesophyll cells allows circulation of air **[1 mark]**; stomata present allow diffusion of gases in and out of the leaf **[1 mark]**

## Reproduction and Inheritance

### Answers to Quick Test Questions

**Page 59**
1. Sexual reproduction involves a male and a female gamete, which must fuse together (fertilisation). Offspring produced by sexual reproduction are not genetically identical to either parent. Asexual reproduction is the production of genetically identical offspring from a single parent. It does not involve fertilisation of gametes.
2. A cell formed by fertilisation (fusion of male and female gametes)
3. Male gametes are made in testes, female gametes are made in ovaries.
4. Nutrients, oxygen, antibodies and waste products are exchanged between the blood of the mother and the blood of the foetus.
5. Amniotic fluid protects the developing foetus from knocks and bumps.

**Page 62**
1. **a)** Anther; (accept stamen) **b)** Ovary
2. Transfer of pollen from anther to stigma
3. **a) Any three from:** Large feathery stigmas; Light smooth tiny pollen grains; Long filaments with anthers that hang outside the flower; Small dull flowers **b) Any three from:** Scented, colourful and large flowers; Nectaries; Spiky or sticky pollen grains; Sticky stigmas
4. The pollen grain grows a pollen tube down the style through the micropyle and into the ovule. The male gamete passes down the pollen tube into the ovule and fuses with the female gamete.
5. **a)** Plumule **b)** Radical **c)** Cotyledons **d)** Testa
6. Water, oxygen, warmth
7. Clones are offspring that are genetically identical to their parent.

**Page 64**
1. **a)** A coiled length of DNA made up of genes, found in the nucleus of plant and animal cells
   **b)** A short section of DNA in a chromosome that codes for the production of a single protein and determines a particular characteristic
2. A pairs with T, C pairs with G.
3. An allele that will always control a characteristic and so appears in the offspring.

**Page 66**
1. Alleles
2. Genotype is the combination of alleles that an individual has for a particular characteristic; phenotype is the visible appearance of the characteristic in the individual.
3. Heterozygous
4. **a)** O **b)** A and B

**Page 68**
1. XY
2. 50%
3. For growth, repair, cloning and asexual reproduction
4. In mitosis, two new cells are formed that are genetically identical to each other and to the parent. The cells formed are diploid. In meiosis, four new cells are formed that are genetically different from each other and from both parents. The cells formed are haploid.
5. Cells produced during meiosis undergo fertilisation. Fertilisation is random, producing different combinations of alleles in the offspring.

**Page 71**
1. Mutations are rare, random changes in the genetic material that can be inherited.

2. Charles Darwin
3. Natural selection is the survival and reproduction of organisms that are best adapted to their environment.
4. Antibiotics encourage the development of populations of bacteria that are resistant to the antibiotic.
5. Darwin's ideas were unpopular because they went against the beliefs of the Church.

### Answers to Exam Practice Questions

AO1 **1. a)** Mitosis

> Spelling is important when two words can easily be confused. You must spell mitosis (and meiosis) correctly in answers like this. An examiner will not know which you mean if you write *meitosis* so you won't get any marks. Glycogen and glucagon are two more words to be especially careful about.

AO2 **b)** A: cytoplasm **[1 mark]** B: chromosome **[1 mark]** C: nucleus **[1 mark]**
AO1 **c)** DNA
AO1 **d)** The two cells formed are genetically identical to each other **[1 mark]** and to the parent **[1 mark]**.
AO1 **e)** They have two sets of chromosomes.
AO1 **f)** 23
AO2 **g)** Four cells produced from each parent cell by meiosis, not two as in mitosis **[1 mark]**; the four cells are genetically different from each other, and from the original cell; whereas in mitosis the cells produced are genetically identical and identical to the parent cell. **[1 mark]**; the gametes produced are haploid, whereas cells formed by mitosis are diploid **[1 mark]**.

AO1 **2. a)** A: anther **[1 mark]** B: stigma **[1 mark]** C: ovary **[1 mark]**
AO2 **b)** By insects
AO2 **c) Accept suitable answer**: e.g. Large colourful petals **[1 mark]**; Attract insects **[1 mark]**

> If the question asks you to write about something shown in the diagram, or in a table, make sure you write about something you can see in that diagram/table. In this question, for example, you would get **no marks** for stating that insect-pollinated flowers have nectaries to attract insects, or that they have scented flowers because no nectaries/scents are shown in the diagram.

AO2 **d i)**
Parent 1 phenotype **Red** X Parent 2 phenotype **Red**
Parent 1 genotype **Rr** X Parent 2 genotype **Rr**
Gametes    **R**   **r**    X **R**   **r**
**[3 marks – one for each correct line]**

| Genotypes | R | r | |
|---|---|---|---|
| R | RR | Rr | Offspring |
| r | rR | rr | |
| | Offspring | | |

**[maximum 2 marks, deduct 1 per error]**
AO1 **ii)** 1 in 4 / 25%

# Answers

## Ecology and the Environment

### Answers to Quick Test Questions

**Page 75**
1. A community is the total number of individuals of all the different populations of plants and animals that live together in a habitat at any one time. A population refers to the number of just a single species in the habitat.
2. **Any three from:** Temperature; Water; Light; Nutrients; Oxygen and Carbon Dioxide
3. Set out a sample area (e.g. 100m$^2$); place 1m$^2$ quadrats randomly in the sample area; count the number of the species in each quadrat; calculate the mean number in each quadrat; multiply this number by the total area of the field.
4. There may be fewer daisies, because the hedge might shade them from light and compete for the water and nutrients in the soil. So the daisies will grow less well.
5. Kite diagrams allow you to compare the pattern of distribution of different species along a transect.

**Page 77**
1. Not all of the Sun's energy is captured by the plant; the consumer uses some of the energy of the producer in movement and respiration and so it is not stored to be passed on; the consumer doesn't eat all of the organism in the trophic level below, e.g. roots, bones, teeth; not all of the energy-containing food consumed is absorbed, some is egested
2. 90%
3. To record biomass the organisms have to be killed and dried out to get rid of the mass of water. This can be very difficult, especially with large organisms like trees.
4. Pyramids of numbers can be inverted if many primary consumers feed on a single larger producer, e.g. a tree.
5. The energy in the organisms found in an area over a period of time

**Page 79**
1. Respiration, combustion, volcanic eruptions
2. Photosynthesis and absorption of carbon dioxide by oceans
3. Approximately 78%
4. Convert ammonium compounds into nitrates

**Page 81**
1. Cause – Burning fossil fuels; Effect – **any one from:** Acidifies lakes and rivers, and kills fish; Corrodes metal; Dissolves buildings; Destroys forests
2. The blood cannot carry as much oxygen.
3. When nutrients are washed out of the soil into lakes and rivers.
4. Peat usually acts as a sponge that soaks up rainwater. If peat is removed, the rainwater runs straight off the land, causing floods.
5. The sewage causes growth of algae. The algae block light to lower plants that will then die because they can't photosynthesise.

### Answers to Exam Practice Questions

AO2 **1. a)** Grass
AO2 **b)** Snake, kookaburra or eagle
AO2 **c)** Wedge-tailed eagle or kookaburra
AO3 **2. a)** Place (ten) 1m$^2$ quadrats randomly in the area **[1 mark]**. In each quadrat count the number of grasshoppers and calculate the mean number per quadrat **[1 mark]**. Multiply the mean number by the total area being studied **[1 mark]**.
AO2 **b)** Some grasshoppers may be counted twice / some grasshoppers may jump out of the quadrat before they are counted **[1 mark]**; so the reliability of the results will be reduced **[1 mark]**.
AO3 **3. a)** There is not enough energy to support another organism in the chain.
AO3 **b)** $\frac{120}{2200}$ × 100 **[1 mark]** = 5.45% **[1 mark]**
AO2 **4.** Acid rain acidifies lakes **[1 mark]**; kills fish **[1 mark]**; so eagles eat more snakes **[1 mark]**; so population of snakes falls **[1 mark]**

> If you are asked to **suggest** an answer, you are expected to use the science you have been taught to work out reasons that explain something. This often happens when the question is about an unfamiliar example of Biology.

AO1 **5. a)** Nitrifying bacteria **[1 mark]**; nitrogen-fixing bacteria **[1 mark]**
AO1 **b) Any two from:** Denitrifying bacteria convert nitrates to nitrogen gas; Nitrates are absorbed from soil by plants; Leaching of nitrate into water
AO2 **c)** Adding fertilisers / compost
AO2 **d)** Excess nitrate causes algae in water to grow and block out sunlight from deeper plants **[1 mark]**; water plants die and are decomposed **[1 mark]**; decomposing bacteria multiply and use up the oxygen in the water **[1 mark]**; oxygen level in the water falls, so the fish die **[1 mark]**

# Answers

## Use of Biological Resources

### Answers to Quick Test Questions

**Page 86**
1. Carbon dioxide; light intensity; temperature.
2. For making amino acids / proteins; for plants to grow.
3. Magnesium is needed for photosynthesis; without photosynthesis the plant cannot grow.
4. **Any three from:** Potassium; Phosphorus; Magnesium; Nitrates
5. A chemical that kills weeds that compete for water, nutrients and light, but does not kill the crop plant.
6. When a predator is introduced that will kill crop pests.

**Page 89**
1. Anaerobic
2. *Lactobacillus*
3. They are added to the boiled and cooled milk where they feed on the milk sugars and produce lactic acid.
4. Food, enzymes, antibiotics, insulin
5. So that no microorganisms are allowed to grow other than the ones you are trying to culture.

**Page 90**
1. Because they are overcrowded
2. The fish are treated with antibiotics and pesticides
3. Antibiotics and pesticides from the tanks can leak into the environment; nutrients from the tanks can escape and cause eutrophication; the tank fish are fed on wild fish meal, so wild fish are still overfished
4. Competition between members of the same species

**Page 91**
1. Lettuces with curly leaves are allowed to flower and are self-pollinated. The seeds are then collected and grown.

**Page 93**
1. Restriction enzymes cut DNA at specific places; ligase enzymes join one piece of DNA to another.
2. A small circular loop of a gene found in bacteria
3. Vectors are used to carry a gene from a cell and insert it into another cell.
4. Fermenters provide the optimum conditions for the growth of genetically modified bacteria that can produce insulin.
5. **Any one from:** The food yield will increase / economic benefit (more food produced for less money); Cheaper because farmers won't need to spray crops with pesticides

**Page 95**
1. Plants can be cloned from tiny pieces of plant tissue that are easy to obtain; whole animals can only be cloned using egg cells that are much more difficult to get. The egg cells have to have their nuclei removed and put into other animal cells, and this process is hard to do and takes a lot of skill. Plant clones are grown in nutrient gel in tubes; animal clones have to be implanted into the uterus of an animal and grow like a normal foetus.
2. The plant clones will all have the same desirable characteristics; tissue culture enables us to mass produce plants easily.
3. **Any two from:** Cloned animals may not be as healthy as normal animals; Cloning reduces genetic variation; Cloned animals are identical copies, so they are all genetically the same – there is potential for one disease wiping them all out

### Answers to Exam Practice Questions

AO2 **1.a)** To keep the microorganisms well mixed with nutrients and oxygen **[1 mark]**; to maintain an even temperature **[1 mark]**
AO2 **b)** To provide the microorganisms with oxygen **[1 mark]** but without contaminating the fermenter **[1 mark]**
AO2 **c)** To provide the microorganisms with sugar **[1 mark]** and protein **[1 mark]**, so that they grow **[1 mark]**
AO2 **d)** So that it is possible to keep the temperature at the optimum **[1 mark]**; too high and the enzymes will denature **[1 mark]**, too low and the microorganisms will grow slowly **[1 mark]**

> When writing about enzymes and temperature, make sure you never say that the enzyme is 'killed'. Remember that enzymes are just chemicals (they are protein molecules) so they are not alive. High temperatures denature enzymes, which means they no longer catalyse reactions.
>
> Use precise terms too: **'denatured'** is much better than **'destroyed'** in this answer.

AO1 **2.a) Accept five suitable points:** The human gene for insulin production is identified; The insulin gene is cut from the DNA using restriction enzymes; The restriction enzyme cuts open bacterial plasmids; Ligase enzymes attach the gene to the plasmid; The plasmids with the insulin gene are taken up by bacteria **[5 marks]**
AO1 **b)** If other microorganisms enter the fermenter they will use up nutrients meant for the modified organisms, which will be expensive **[1 mark]**; Some contaminating bacteria may be harmful to humans **[1 mark]**
AO2 **c)** Insulin can be mass produced economically **[1 mark]**; GM insulin causes fewer side effects than insulin obtained from pigs **[1 mark]**
AO1 **3.a) Any two from:** Farmers will produce more food because the crop won't be damaged by pests; It will be cheaper because the farmer doesn't have to spray the crop with pesticide
AO1 **b)** People are afraid that the genetic modification will spread to other plants with unknown effects **[1 mark]**; People worry that GM crops are harmful to eat **[1 mark]**
AO3 **4.a)** Food for yeast **[1 mark]**; so yeast could grow; **[1 mark]**; for yeast respiration **[1 mark]**
AO3 **b)** Time taken for the X to disappear **[1 mark]**
AO3 **c) Any two from:** The amount of yeast added; The concentration of glucose; The volume of solution used; The distance of the card from the flask **[2 marks]**
AO3 **d)** The yeast reproduced **[1 mark]** and made the liquid too cloudy to see through **[1 mark]**
AO3 **e)** 40°C **[1 mark]**
AO3 **f)** Too hot **[1 mark]**. Yeast enzymes denatured **[1 mark]**
AO3 **g)** Anaerobic **[1 mark]**
AO2 **h)** Alcohol **[1 mark]**
AO3 **i) Accept three from:** Repeat the experiment at more temperatures; Repeat the experiments at each temperature and take an average time; Take the readings at shorter time intervals; Set up controls with boiled and cooled yeast, for each temperature **[3 marks]**

> There are often more answers possible for a question, than marks available. Use the mark allocation as a guide, but be careful as marks are given for separate points, not for saying the same point in two different ways.

# Glossary of Key Words

**Absorption** – substances are taken into the organism usually by diffusion, osmosis or active transport

**Active site** – a 3D space in the surface of an enzyme molecule, which the substrate fits into

**Active transport** – movement of substances across a membrane against a concentration gradient, using energy from respiration

**Allele** – a form of a gene

**Aerobic respiration** – releases energy inside living cells by breaking down glucose and combining the products with oxygen

**Alveoli** – microscopic air sacs in the lungs where gas exchange takes place

**Anaerobic respiration** – releases energy inside the cytoplasm of living cells by breaking down glucose molecules without the use of oxygen

**Antibiotic** – a chemical used to kill or disable bacteria that cause disease

**Antibody** – chemical produced by lymphocytes which acts against pathogens that infect the body

**Anti-diuretic hormone (ADH)** – a hormone that regulates the water content of the blood. It increases the reabsorption of water in the kidney

**Antigen** – a protein molecule that the immune system recognises as foreign (not part of the body)

**Artery** – carries blood away from the heart towards the organs

**Asexual reproduction** – production of a new individual from a parent cell, without the need for fertilisation

**Assimilation** – the process through which complex molecules used in the body are built up from simple nutrient molecules that have been absorbed

**Atheroma** – fatty material found in arteries

**Auxin** – plant hormone

**Bacteria** – a large group of very simple single-celled organisms, some of which cause disease

**Benedict's Reagent** – used to test foods for glucose

**Bile** – produced in the liver and stored in the gall bladder; neutralises the acid added to food in the stomach; emulsifies lipids

**Biodiversity** – the variety of life

**Biomass** – the mass of living material

**Bronchi** – branches of the trachea

**Bronchioles** – branches of a bronchus

**Capillary** – very narrow vessel that carries blood between arteries and veins, and through the organs

**Carbohydrate** – a large nutrient molecule made of sugars joined together, e.g. starch and glycogen

**Carcinogen** – cancer-causing chemical

**Cardiovascular disease** – disease of the heart and blood vessels

**Catalase** – an enzyme that breaks down hydrogen peroxide into oxygen and water

**Catalyst** – a chemical substance that speeds up a chemical reaction but is not used up or changed by the reaction

**Cell membrane** – structure that surrounds the cytoplasm and controls the passage of substances in and out of the cell

**Cell wall** – a tough layer that surrounds plant cells, fungal and bacterial cells

**Central nervous system** – brain and spinal cord

**Chemical digestion** – the breakdown (using chemicals) of large molecules into small ones that can be absorbed

**Chitin** – what the cell walls of hyphae are made from

**Chlorophyll** – green pigment found in chloroplasts, contains magnesium and traps light energy during photosynthesis

**Chloroplast** – plant cell structure that contains chlorophyll and is where photosynthesis takes place

**Chromosome** – a long coiled molecule of DNA, divided up into regions called genes, also made of DNA

**Clone** – cells or organisms that are genetically identical; plant clones are grown from cuttings or tissue culture and are genetically identical to the parent plant

**Community** – the total number of individuals of all the different populations of plants and animals that live together in a habitat at any one time

**Concentration gradient** – difference in concentration across a membrane

**Consumer** – an organism which eats other organisms

**Cotyledon** – embryo leaf, where food is stored

**Cytoplasm** – where most chemical reactions take place in plant cells

**Decomposer** – organism, e.g. fungus or bacterium, that feeds on dead material and causes it to decay

**Deforestation** – the large-scale cutting down of trees for timber and land

**Denatured** – changed irreversibly

**Deoxygenated** – blood that oxygen has been released from

**Dependent variable** – the variable that is measured during an experiment

**Differentiate** – become specialised

**Diffusion** – the movement of a substance from a region of high concentration to a region of low concentration

**Digestion** – breaking down large, insoluble nutrient molecules into small, soluble ones that can be absorbed

**Dominant allele** – the form of the gene which shows as a characteristic in the individual

# Glossary of Key Words

**Double helix** – two helix shaped molecules that are twisted and bonded together (e.g. DNA)

**Ecosystem** – a physical environment with a particular set of conditions, plus all the organisms that live in it

**Effector** – organ or cell that acts in response to a stimulus

**Egestion** – when undigested food and excess water leave the body through the anus

**Embryo** – developing young organisms where the cells are becoming specialised and organised into the organs and systems of the body

**Emphysema** – lung disease that enlarges and damages the alveoli

**Enzyme** – biological catalyst made of protein – enzymes speed up metabolic reactions

**Excretion** – removal of waste products of metabolism from an organism

**Expiration** – breathing out

**Extinct** – when a species has died out

**Extra-cellular digestion** – when fungi secrete digestive enzymes onto food

**Fermentation** – anaerobic respiration in yeast

**Fertilisation** – when male and female gametes fuse together during sexual reproduction

**Fertiliser** – substances added to soil that increase the mineral levels

**Gamete** – a sex cell, i.e. female egg or male sperm

**Gene** – a section of DNA that codes for a particular protein

**Genetic modification** – the addition of genetic material to a cell or an organism that gives the cell/organism a new characteristic, e.g. the introduction of a gene that gives a plant resistance to herbicides

**Genotype** – the combination of alleles that an individual has for a particular gene

**Global warming** – increase in temperature of the Earth's surface and atmosphere, and that may be due to increased greenhouse effect

**Glycogen** – a carbohydrate used as an energy store in animals

**Greenhouse effect** – warming of Earth's atmosphere that results from increased levels of greenhouses gases, which prevent heat being radiated from the Earth into Space

**Growth** – permanently increasing mass – cells increase in number and size

**Habitat** – part of the physical environment where an animal or plant lives

**Haemoglobin** – chemical found in red blood cells, which easily attaches to oxygen and transports oxygen around the body

**Heterozygous** – an individual who carries two different alleles for a particular gene

**Homeostasis** – the maintenance of a constant internal environment in the body, e.g. temperature

**Homozygous** – an individual who carries two copies of the same allele for a particular gene

**Hormone** – a chemical message released by a gland directly into the bloodstream

**Hyphae** – fine threads that make up fungi

**Hypothalamus** – part of the brain that monitors how much water is present in the blood

**Immune system** – system which protects the body against infection

**Independent variable** – the variable that the scientist alters during an investigation

**Ingestion** – when you put food or drink into your mouth.

**Inspiration** – breathing in

**Interspecific predation** – when a member of one species eats a member of a different species

**Intraspecific predation** – when a member of one species eats a member of the same species

**Iodine** – used to test foods for starch

**Leaching** – when mineral ions dissolve in soil water and are washed out of the soil into lakes and rivers

**Ligase enzyme** – enzyme involved in joining pieces of DNA together

**Lipid** – a large nutrient molecule built from fatty acids and glycerol

**Mechanical digestion** – where teeth, muscles and bile break up lumps of food into a fine mush that has a large surface area for enzymes to act on

**Meiosis** – a type of cell division which occurs in the testes and ovaries

**Menstrual cycle** – regular sequence of events in a female reproductive system that prepare the uterus to receive a fertilised egg. (Without fertilisation the uterus lining is shed – menstruation.)

**Metabolism** – chemical reactions inside the body that keep life processes working

**Microorganism** – microscopic, usually single-celled, organism

**Micropyle** – pore through which pollen tube enters ovule

**Mitochondria** – where most energy is released in respiration in a cell

**Mitosis** – type of cell division that produces two genetically identical daughter cells, each of which is also genetically identical to the original cell

# Glossary of Key Words

**Movement** – one of a plant or animal's response to extended stimuli

**Mutation** – rare, random change in genetic material that can be inherited

**Mycelium** – mass of hyphae that forms the body of a fungus

**Natural immunity** – immune response that occurs when an organism infects the body

**Natural selection** – the survival and reproduction of organisms that are best adapted to their environment

**Neurone** – part of the nervous system that carries nerve impulses between the peripheral and central nervous systems

**Nucleus** – controls the activities of a cell

**Nutrition** – taking in and using nutrients as raw materials to build cells and to release energy

**Oestrogen** – a hormone that stimulates the uterus lining to thicken

**Optimum temperature** – temperature at which enzymes work fastest

**Organ** – made of tissues

**Organelle** – structure within a cell that carries out a particular function, e.g. nucleus, chloroplast

**Osmoregulation** – regulating water and salt content of a cell or of the blood

**Osmosis** – the diffusion of water from a high concentration of water (dilute solution) to a low concentration of water (concentrated solution) through a partially permeable membrane (a membrane that allows the passage of water molecules but not solute molecules)

**Ovule** – female structure in a flower that contains the female gamete (egg cell)

**Oxygenated** – when blood is loaded with oxygen

**Oxyhaemoglobin** – molecule of haemoglobin attached to oxygen molecules

**Pacemaker** – part of the heart that triggers each beat

**Pasteurise** – heating to kill bacteria, then cooling

**Pathogen** – an organism that can cause disease

**Peat** – a partially decayed vegetation used for fuel and fertilisers by gardeners

**Peristalsis** – the squeezing action of the gut that pushes food along

**Pest** – organism that harms a cultivated or farmed organism such as a crop plant or livestock

**Pesticide** – used to kill pests that damage crops or livestock so that more food is produced

**Phenotype** – the expression of the genotype (the characteristic shown)

**Phloem** – transports substances around a plant

**Photosynthesis** – the process in which green plants make their food, using sunlight

**Pituitary gland** – important endocrine gland, found in the brain

**Plasmid** – tiny loops of DNA; all bacteria have plasmids

**Plumule** – embryo shoot

**Pollen** – male structure produced by anthers, contains the male gamete

**Pollen tube** – structure that grows from the pollen grain down the style to the ovule; male gamete travels down it to reach the ovule

**Pollution** – damage to the environment by chemicals added to land, water or air

**Population** – the total number of individuals of the same species that live in a certain area

**Potometer** – apparatus used to measure transpiration rate

**Predator** – organism that feeds on another organism

**Producer** – organism that produces its own food by photosynthesis, and so converts light energy from the Sun into chemical energy in carbohydrates

**Progesterone** – hormone produced by the ovary to preserve the uterus lining

**Protein** – a large nutrient molecule made of amino acids joined together

**Pyramid of biomass** – shows the dry mass of living material at each stage in the food chain

**Pyramid of numbers** – shows the number of organisms at each stage in a food chain

**Quadrat** – square frame, half or one metre square usually, that is used to sample the organisms in an area

**Quantitative** – using measured or calculated numerical values

**Radical** – embryo root

**Receptor** – structure that detects a stimulus and converts it into a nerve impulse

**Recessive allele** – allele that codes for a characteristic but that is masked if the dominant allele is also present

**Reflex action** – nervous system response that occurs without conscious control

**Reproduction** – creating new members of the species

**Respiration** – the release of energy from food chemicals in all living cells

**Respiratory system** – the organs involved in ventilation of the gas exchange surface in animals

# Glossary of Key Words

**Restriction enzyme** – enzyme used in genetic engineering to cut DNA

**Retina** – part of the eye that contains light sensitive cells

**Ribosome** – where protein synthesis takes place in a plant cell

**Root hair** – root hairs project from the cells of the root tip. They have an enormous surface area for absorbing water and so increase the plant's ability to take up water

**Saprotrophic nutrition** – when fungi secrete digestive enzymes onto food and then absorb the products of digestion

**Selective breeding** – when plants or animals with certain traits are reproduced to produce offspring with certain desirable characteristics

**Sensitivity** – detection of changes (stimuli) in the surroundings and the ability to respond to them

**Sexual reproduction** – production of new individuals that involves the fusion of male and female gametes

**Soil erosion** – removal of soil by flooding, rainfall and wind

**Stem cell** – cell that has the ability to divide and specialise into any type of cell

**Stimulus** – a change in the environment

**Stomata** – tiny pores on the underside of a leaf, to allow the exchange of gases

**Testa** – seed coat

**Tissue** – a group of cells that have a similar structure and function

**Trachea** – a flexible tube, surrounded by rings of cartilage to stop it collapsing

**Transect** – a measure line across a habitat, along which the distribution of different species is recorded

**Transgenic** – animals and plants whose cells have been modified by the transfer of genetic material from a different species

**Transpiration** – the loss of water vapour from a leaf, by evaporation from cell surfaces and diffusion through the stomata

**Trophic level** – the position or stage that an organism occupies in a food chain

**Ultrafiltration** – filtration of the blood under high pressure that occurs in the glomerulus, and that removes all small molecules and ions from the blood into the Bowman's Capsule

**Vacuole** – large cell structure containing cell sap

**Variation** – differences between individuals of the same species

**Variegated leaf** – only has chlorophyll in the parts that are green

**Vector** – structure that carries genetic material into a cell that is being genetically modified; can be a plasmid or a virus

**Vein** – carries blood from the organs back to the heart

**Ventilation** – movement of air in and out of the lungs

**Villi** – microscopic finger-like projections that project from cells lining the small intestine

**Virus** – viruses are not cells; they are DNA or RNA surrounded by protein. Cause diseases

**Xylem** – transports substances around a plant

**Yeast** – a single-celled fungus that reproduces by budding

**Zygote** – a cell formed by fertilisation (fusion of male and female gametes)

---

**P2** **Amniotic fluid** – protects the developing foetus from knocks and bumps

**Co-dominant** – when both the alleles are equally dominant, so characteristics of both alleles are present in the organism

**Control** – an experiment carried out to check that the results recorded from a test experiment are only due to the changes in an independent variable, not something else.

**Flaccid** – not rigid

**Foetus** – an unborn human baby / animal baby

**Mycoprotein** – a protein-rich food suitable for vegetarians made by using Fusarium, a type of fungus

**Nitrate** – soluble compound that contains nitrogen in a form that plant roots can absorb

**Obese** – 20% above ideal weight

**Optimum pH** – the pH level at which an enzyme works best

**Placenta** – structure formed by the foetus, that attaches to the uterus wall, and is where exchanges take place between the mother's and the foetus's blood.

**Plasmolysis** – when cells lose a lot of water and the inside of the cells contract

**Translocation** – the movement of food substances (sucrose and amino acids) around a plant

**Turgid** – rigid

**Umbilical cord** – how a developing foetus is joined to the placenta

# Notes

# Notes

# Index

## A
Absorption 25
Active site 14
Active transport 16
Aerobic respiration 28
Alleles 64
Alveoli 32
Amniotic fluid 59
Amylase 15
Anaerobic respiration 28
Animals 5
Antibiotic 43
Antibody 43
Anti-diuretic hormone (ADH) 47
Antigen 44
Artery 41
Artificial controls 84
Asexual reproduction 58
Assimilation 25
Atheroma 34
Atria 42
Auxin 50

## B
Benedict's Reagent 13
Bile 11, 25
Biodiversity 80
Biomass 76
Blood 40
Breathing 33
Brewing beer 87
Bronchi 32
Bronchioles 32

## C
Capillary 41
Carbohydrate 13
Carbon cycle 78
Carcinogen 34
Cardiac cycle 42
Cardiovascular disease 34
Catalase 15
Catalyst 13
Cell membrane 5, 10
Cell structures 10
Cell wall 5, 10
Central nervous system 51
Chemical digestion 25
Chitin 6
Chlorophyll 19
Chloroplast 5, 10
Chromosomes 63
Clone 62
Cloning 94–95
Co-dominant 65
Community 74
Concentration gradient 16, 17
Cotyledon 61
Cytoplasm 5, 10

## D
Decomposer 76
Deforestation 80
Denatured 14
Deoxygenated 42
Dependent variable 15
Differentiate 11
Diffusion 16
Digestion 25
Digestive system, 11, 24
DNA 63
Dominant allele 64
Double helix 63

## E
Ecosystem 74
Effector 49
Egestion 25
Embryo 58
Emphysema 34
Enzyme 6, 13, 14
Essential minerals 21
Excretion 4, 45–47
Expiration 33
Extinction 70
Extra-cellular digestion 6
Eye 52

## F
Fermentation 87
Fertilisation 58, 64
Fertilisers 21, 85
Fish farming 90
Foetus 59
Fungi 6

## G
Gametes 64
Gene 63
Genetic modification 92
Genotype 66
Global warming 80
Glycogen 6
Greenhouse effect 80
Growth 4

## H
Habitat 74
Haemoglobin 40
Heart 42
Heterozygous 66
Homeostasis 4
Homozygous 66
Hormones 49
Hyphae 6
Hypothalamus 47

## I
Immune system 43
Independent variable 15
Indigestion 25
Industrial fermenters 89
Inspiration 33
Intraspecific predation 90
Interspecific predation 90
Iodine 13

## K
Kidneys 45
Kingdoms 5

## L
Large intestine 11
Leaching 81
Ligase enzyme 92
Lipid 13
Liver 11

## M
Making yoghurt 89
Mechanical digestion 25
Meiosis 68
Menstrual cycle 59
Metabolism 45
Micropyle 60
Mitochondria 5, 10
Mitosis 64, 68
Monohybrid inheritance 66
Movement 4
Multicellular 5
Mutation 69
Mycelium 6

## N
Natural immunity 44
Natural selection 70
Net movement 16
Neurone 51, 52
Nitrogen cycle 79
Nitrates 79
Nucleus 5, 10
Nutrition 4

## O
Obesity 23
Oestrogen 59
Optimum pH 14
Optimum temperature 14
Organ systems 11
Organelle 10
Osmoregulation 45
Osmosis 16, 18
Ovule 60
Oxygenated 42
Oxyhaemoglobin 40

## P
Pacemaker 42
Pancreas 11
Pasteurise 87
Pathogen 7
Peat 81
Peristalsis 24
Permanent vacuole 10
Pest control 85
Pesticides 85
Phenotype 66
Phloem 12
Photosynthesis 19, 20, 21, 22, 30
Pituitary gland 47
Placenta 59
Plant leaves 20
Plant organs 12
Plants 5
Plasmid 6
Plasmolysis 17
Plumule 61
Pollen 60
Pollination 60
Pollution 81
Population 74
Potometer 37, 38
Predator 86
Producer 76
Progesterone 59
Protein 13
Protoctists 7
Pyramid of biomass 77
Pyramid of numbers 77

## Q
Quadrat 74
Quantitative 74

## R
Radical 61
Receptor 49
Recessive allele 64
Recycling 78
Reflex action 52
Reproduction 4, 58
Respiration 4, 28–29
Respiratory system 33
Restriction enzyme 92
Retina 52
Ribosome 5, 10
Root hair 36

## S
Salivary glands 11
Saprotrophic nutrition 6
Selective breeding 91
Sensitivity 4
Sexual reproduction 58
Small intestines 11
Soil erosion 81
Specialised cells 10
Stem cells 95
Stimulus 48
Stomach 11, 20, 30

## T
Testa 61
Thorax 32
Tissues 11
Trachea 32
Transect 75
Transgenic 92
Translocation 36
Transpiration 37, 38
Trophic level 76

## U
Ultrafiltration 46
Umbilical cord 59

## V
Vacuole 5
Variation 69
Variegated leaf 21
Vector 92
Ventilation 33
Villi 26
Viruses 5, 7

## W
Water cycle 78
Water pollution 81

## X
Xylem 12

## Y
Yeast 87

## Z
Zygote 58